The Shortest-Path Problem
Analysis and Comparison of Methods

Synthesis Lectures on Theoretical Computer Science

The Shortest-Path Problem: Analysis and Comparison of Methods
Hector Ortega-Arranz, Diego R. Llanos, and Arturo Gonzalez-Escribano
2014

The Shortest-Path Problem: Analysis and Comparison of Methods

Hector Ortega-Arranz, Diego R. Llanos, and Arturo Gonzalez-Escribano

ISBN: 978-3-031-01446-8   paperback
ISBN: 978-3-031-02574-7   ebook

DOI 10.1007/978-3-031-02574-7

A Publication in the Springer series
SYNTHESIS LECTURES ON THEORETICAL COMPUTER SCIENCE

Lecture #1
Series ISSN
ISSN pending

The Shortest-Path Problem

Analysis and Comparison of Methods

Hector Ortega-Arranz, Diego R. Llanos, and Arturo Gonzalez-Escribano
Universidad de Valladolid, Spain

SYNTHESIS LECTURES ON THEORETICAL COMPUTER SCIENCE #1

ABSTRACT

Many applications in different domains need to calculate the shortest-path between two points in a graph. In this paper we describe this shortest path problem in detail, starting with the classic Dijkstra's algorithm and moving to more advanced solutions that are currently applied to road network routing, including the use of heuristics and precomputation techniques. Since several of these improvements involve subtle changes to the search space, it may be difficult to appreciate their benefits in terms of time or space requirements. To make methods more comprehensive and to facilitate their comparison, this book presents a single case study that serves as a common benchmark. The paper also compares the search spaces explored by the methods described, both from a quantitative and qualitative point of view, and including an analysis of the number of reached and settled nodes by different methods for a particular topology.

KEYWORDS

graph algorithms, graph search, shortest-path problem, Dijkstra's algorith, A* algorithm, highway hierarchies, contraction hierarchies, transit-node routing, hub-based labeling algorithm, landmark A*, edge flags, reach-based routing, precomputed cluster distances

Contents

List of Figures

List of Tables

Acknowledgments

This research is partly supported by the Ministerio de Industria, Spain (CENIT OCEAN-LIDER), MINECO (Spain) and the European Union FEDER (CAPAP-H networks TIN2010-12011-E and TIN2011-15734-E; TIN2011-25639), Castilla y León regional government (VA172A12-2), and the HPC-EUROPA2 project (project number: 228398) with the support of the European Commission - Capacities Area - Research Infrastructures Initiative. We also would like to thank GMV for its support, and the anonymous referees for their comments.

Hector Ortega-Arranz, Diego R. Llanos, and Arturo Gonzalez-Escribano
December 2014

CHAPTER 1

Introduction

To find out the shortest path between an origin and a destination in a graph is an important problem whose solution has many applications, including car navigation systems [Sanders et al., 2008], traffic simulations [Barceló et al., 2005], scheduled means of transport [Bast et al., 2010, Bauer et al., 2007], logistic control [Chen and Lee, 2011], spatial databases [Papadias et al., 2003, Shekhar et al., 1997], Internet route planners [Rétvári et al., 2007], or web searching [Barrett et al., 2000, Böhm et al., 2011]. Despite the importance of the Shortest-Path Problem (SPP), algorithms to solve it are computationally costly, and in many cases commercial products implement heuristic approaches to generate approximate solutions instead. Although heuristics are usually faster and do not need much amount of data storage or precomputation, they do not guarantee the optimal route.

The solution to the problem can be divided in two phases. Given an input graph, the *query phase* is the search of the shortest path from a source location to a destination. During the last decade, many techniques have been presented to reduce the query time. Usually, this goal is achieved using additional information returned from a previous processing of the original data. This previous step is known as *precomputation phase*. Precomputed data is the basis of the speed-up achieved by modern approaches. However, the generation and use of precomputed data depend strongly on the application domain. In a static scenario, where the graph information does not change frequently, like road networks, precomputation step is only carried out from time to time, so higher preprocessing times are usually affordable. Problems arise when it is desired to enrich the data, for example adding dynamic traffic status or road maintenance information. Frequent changes in the data require to run the precomputed phase more frequently, quickly becoming a performance bottleneck. Other situations that require re-running the precomputation phase include changes of the metric used to measure the shortest path, such as travel time, energy saving considerations, or the use of different means of transport. It is prohibitive to deploy the precomputation process every time the road network model changes. In fact, there are some techniques that present a good balance between flexibility, precomputation, and query times, by just updating the part of the data involved in the change, instead of precomputing again all the information every time a change occurs.

The advent of mobile devices present a third challenge, not only due to their limited computing power, but also because of their small memory size, that limits the amount of precomputed data that can be stored. Most of the current approaches present a trade-off between the amount of memory used and the query time needed: The more memory used, the better the query time.

Therefore, current solutions can be customized depending on the particular hardware requirements.

As we will see later, huge speed-ups have been achieved during the last years in the query phase [Bast et al., 2014, Sommer, 2014]. With query times performed in the order of nanoseconds [Abraham et al., 2012, Delling et al., 2013a], almost no room for improvement is left. However, the use of parallel techniques may transform otherwise prohibitive methods in feasible solutions, reducing data space consumption, and query or precomputation times.

This work is organized as follows. Chapter 2 shows a brief summary of the graph theory needed to understand the solutions to the SPP problem, and presents a case study that will later be used as an example to show the behavior of each method described. Chapter 3 describes some classical approaches to this problem: Dijkstra's algorithm together with some refinements in the definition and use of its data structures; the Bidirectional Search; and heuristic-based approaches such as A* and the Bidirectional A*. Chapter 4 describes some methods that use the preprocessing step to augment the graph with a hierarchical classification of graph edges. Chapter 5 shows some non-hierarchical methods that reduce the exploration space using information obtained in a preprocessing step. We also cite some efforts that combine hierarchical with non-hierarchical techniques. Chapter 6 compares all methods in terms of their search space, both quantitatively and qualitatively. Finally, Chapter 7 concludes our work.

CHAPTER 2

Graph Theory Basics

The easiest way to represent the information of a network is to transform it into a graph, where every link will be an edge, and every possible joint to change to another link will be a node. Some groups of nodes could represent cities, stations, or only intersection points, depending on the scenario to be considered. In this chapter, we present a summary of graph theory, focused on the concepts related to the shortest-path problem. More general introductions to this topic can be found in classical books such as Bondy and Murty [1976] or West et al. [2001]. We will also present a graph that will later be used as a basis to explain all the algorithms covered in this work.

2.1 DEFINITIONS

A graph $G = (V, E)$ is composed by a set of vertices V, also called nodes, and a set of edges E, also called arcs. Every node v is usually depicted as a point in the graph. Every edge e is usually depicted as a line that connects two and only two nodes. An edge is a tuple (u, v) that represents a link between nodes u and v. The number of edges connected to a node v is called the *degree* of v. In an *undirected graph* all edges can be traversed in both directions, whereas an edge (u, v) of a directed graph only can be traversed from u to v. Associated to each edge there is a weight function $w(u, v)$, that represents the cost of traversing this edge.

A *path* $P = \langle s, ..., u, ..., v, ..., t \rangle$ is a sequence of nodes connected by edges, from a source node s to a target one t (see Figure 2.1). A path $P' = \langle u, ..., v \rangle$ is a *subpath* of P if its sequence of nodes and edges are contained in the same order in P. The *weight* of a path, $w(P)$, is the sum of all the weights associated to the edges involved in the path. The shortest path between two nodes s and t is the path with the minimum weight among all possible paths between s and t. Finally, the minimum distance between s and t, $d(s, t)$, is the weight of the shortest path between them.

2.2 A STUDY CASE

Figure 2.2 represents a synthetic graph designed to serve as an example to evaluate the strengths and weaknesses of the approaches described in the following sections. It will also be used to show the steps deployed in preprocessing and query phases. All the approaches studied will be used to solve a query to find the shortest path from source node s to target node t, highlighted in light gray in Figure 2.2, whose weight is $w(P) = 32$. As we will see, each method will apply different criteria to find this sequence of nodes and edges. We will use an undirected graph in order to simplify the explanations of the processes. The graph includes some dotted lines, meaning that

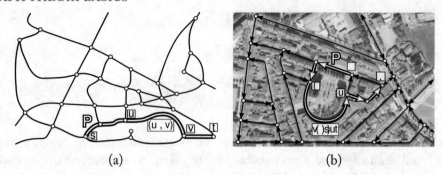

Figure 2.1: Examples of paths in (a) an undirected graph and (b) a directed graph.

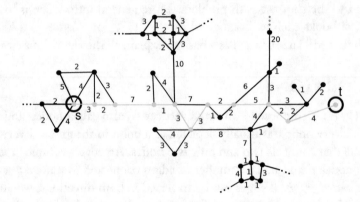

Figure 2.2: Graph example and shortest path from source node s to target node t.

the graph continues in what we call *dark areas*. Unless otherwise noted, to know exactly what is in these areas is not important.

In general, there is a correlation between the number of nodes visited during the search and the running time needed by each algorithm. However, the use of heuristics or pruning criteria implies additional computational costs, that in small graphs may be higher than the time saved by these methods. Our study case will only show the search space visited by each approach.

All search spaces generated by each approach will be compared against the search space generated by Dijkstra's algorithm. Although this comparison might show that there is only a slight difference between them using our case study, these small differences become crucial in bigger graphs.

CHAPTER 3

Classical Algorithms

In this chapter we will review some classical solutions to the shortest-path problem, including Dijkstra's algorithm, some improvements on its data structures, and its bidirectional and heuristic variants. None of these solutions require precomputation. To better show how each algorithm works, we will see how they explore the graph proposed as our case study.

3.1 DIJKSTRA'S ALGORITHM

This algorithm is the basic solution for the shortest-path problem. Dijkstra's algorithm [Dijkstra, 1959] constructs minimal paths from a source node s to the remaining nodes, exploring adjacent nodes following a proximity criterion until the shortest path to goal node t is found. The exploring process is known as *edge relaxation*. When an edge (u, v) is relaxed from a node u, it is said that node v has been *reached*. Therefore, there is a path from source through u to reach v with a tentative shortest distance. Node v will be considered *settled* when the algorithm has found the shortest path from source node s to v. The algorithm finishes when t is settled.

The algorithm uses an array, D, that stores all tentative distances found from source node s to the rest of nodes. At the beginning of the algorithm, every node is unreached and no distances are known, so $D[i] = \infty$ for all nodes i, except current source node $D[s] = 0$.

The algorithm proceeds as follows:

1. (Initialization) It starts on the source node s, initializing distance array $D[i] = \infty$ for all nodes i and $D[s] = 0$. Node s is settled and considered as the *current node* c ($c \leftarrow s$).

2. (Edge relaxation) For every node v adjacent to c that has not been settled, a new distance from source is found using the path through c, with value $D[c] + w(c, v)$. If this distance is lower than previous value $D[v]$, then $D[v] \leftarrow D[c] + w(c, v)$.

3. (Settlement) The node u with the lowest value in D is taken as the new current node ($c \leftarrow u$). After this, current node c is now considered as settled.

4. (Termination criteria) If u is the target node the algorithm finishes, since the tentative path to u is now considered the minimum one. Otherwise, the algorithm proceeds to step 2.

In order to recover the path, every node reached stores its predecessor, so at the end of the query phase the algorithm just runs back from target through stored predecessors till the source is reached. The *shortest-path tree* of a graph from source node s is the composition of every shortest path from s to the remaining nodes. See Figure 3.1.

Regarding the complexity of Dijkstra's approach, let n be the number of nodes, and m the number of edges. In Dijkstra's algorithm, the initialization has a cost in $O(n)$. The updating operation, that has a constant cost, is performed m times in the worst case. The linear search in D for the node u with the lowest value (step 3) is performed in $O(n)$. Since it should be done n times in the worst case, this leads to a total upper bound of $O(n + m + n^2) \in O(n^2)$ [Cormen et al., 2001].

Many authors use the number of settled nodes as a comparing metric. In this context, *efficiency* can be defined as the number of nodes on the shortest path divided by the number of nodes settled by the algorithm [Goldberg and Harrelson, 2004]. With this criteria, Dijkstra's algorithm settled nodes upper bound is in $O(n)$, because the worst case is a query that would visit all the nodes of the graph.

3.1.1   USING PRIORITY QUEUES

As we said above, Dijkstra's solution relies on a vector to store the tentative distances of the reached nodes. For sparse graphs (graphs with a *small* ratio of edges per node), the naïve implementation of Dijkstra's algorithm can be improved using a priority queue to store reached nodes.

To implement a priority queue, the following operations should be defined:

Insert with priority The structure will be sorted with respect to a key, called the priority. The insertions of new elements are done according with their priority, to keep the structure sorted in non-decreasing order. In our case, the key associated to each node is the total distance found so far from the source node.

Update This operation changes the value of the key of an element and reorders the queue.

Find minimum element This operation finds the element of the structure with the lowest priority (key), in our case the lowest distance.

Delete minimum element This operation, also called *pop*, deletes the element from the structure with the lowest priority.

The use of a priority queue helps to reduce the asymptotical behavior of Dijkstra's algorithm. If traditional binary heaps such as Williams's heaps [Williams, 1964] are used as the structure of the priority queue, each inserting, deleting, or updating operation takes $O(\log n)$. Selecting the next node to be processed (the current node) simply implies finding the minimum element, with cost in $O(1)$, and to delete it, with cost $O(\log n)$. The operations described above lead to an execution time in $O((m + n) \log n) = O(m \log n)$. The Fredman and Tarjan's Fibonacci heaps [Fredman and Tarjan, 1987] reduced the running time to $O(n \log n + m)$. For a thorough review of the use of priority queues see Meyer [2001], Thorup [1999].

Besides these improvements in terms of asymptotical complexity, the use of priority queues does not lead to significant performance improvements when processing real-world large road

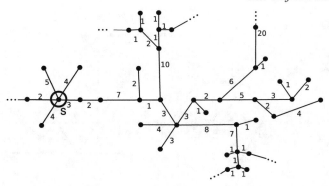

Figure 3.1: Shortest-path tree from source node s.

networks, because the occurrence of cache faults when accessing the graph [Delling et al., 2009] diminishes the returns offered by this technique.

In fact, for such problems, approaches that reduce the nodes to be considered (that will be described later) are much more powerful.

3.1.2   CASE STUDY

In this section we will apply Dijkstra's algorithm using priority queues to our study case. The algorithm starts with the source node s as current node c (see Figure 3.2(a)). By relaxing its edges, tentative paths from the current node to several reached nodes are found. These nodes are inserted into the non-decreasing sorted queue using the distance from the source node as the key. This leads to the following queue: $[(u, 2), (x, 3), (w, 4), (y, 4), (v, 5)]$. All related path distances are updated.

After this phase, the first node in the queue is chosen as the new current node, in our case u, and it is considered settled. By relaxing its edges, we find node s (that was already settled), an already reached node y, and an unreached node from the dark area (see Figure 3.2(b)). For already settled nodes there are no actions. For reached nodes, the algorithm compares if the new tentative path distance to them is lower than the one stored in the queue, in order to update the tentative distance with the new value if necessary. This is not our case. Finally, the unreached node is inserted into the queue in the corresponding place, preserving the order.

The algorithm continues by taking the first node of the queue and settling it, repeating the process until the solution is found. Figure 3.2 (c) shows the order in which the nodes are settled after nine steps (square-bounded numbers), without taking into account the nodes of the dark area at the left. By the time the ninth node is settled, reached nodes j, k, and p have already been inserted in the queue due to the relaxation of the eighth settled node. The queue at this point is $[(k, 16), (j, 20), (p, 23)]$.

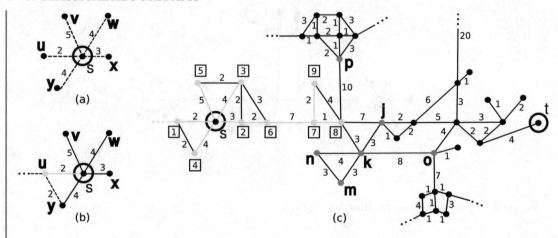

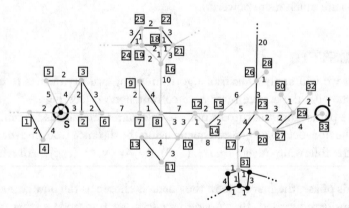

Figure 3.2: First steps of Dijkstra's algorithm. (a) Starting node. (b) Following node to be considered. Solid edges are paths to reached nodes. (c) Partial settling order (square-bounded numbers) after nine steps. Light gray arcs mark the shortest paths from source node s to settled nodes found so far.

Figure 3.3: Settled nodes and Dijkstra's rank.

The algorithm then proceeds by taking the node k (see Figure 3.2 (c)), reaching new nodes n, m, and o, and updating the tentative distance of node j. After these steps, the new queue is $[(j, 19), (m, 19), (n, 20), (p, 23), (o, 24)]$ (note that nodes m and n were inserted between nodes j and p in the priority queue). The algorithm will finish when the settled node is the target.

The settled nodes of Dijkstra's algorithm for the study case is shown in Figure 3.3, with settling order represented by square-bounded numbers. The settling order is known as *Dijkstra's rank*. As can be seen, almost all the nodes of the graph have been settled, together with *every* node x of the dark areas whose distance from source node s, $d(s, x)$, is not higher than $d(s, t)$.

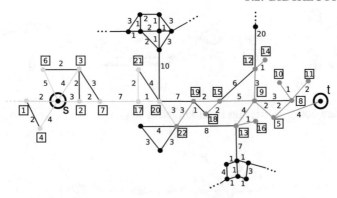

Figure 3.4: Settled nodes and Dijkstra's rank in a bidirectional query.

3.2   BIDIRECTIONAL SEARCH

The bidirectional search algorithm [Dantzig, 1963, Dreyfus, 1967, Nicholson, 1966] alternates between two Dijkstra's searches, one from s to t, and a second one, called backward search, from t to s. In the backward search, the key of a node v is the distance from v to the target node t. Note that, in a directed graph, the reversed edges should be considered for the backward search.

Having two searches to perform, it is possible to maintain two separate queues and use a turn policy to take the next node, or to put all the reached nodes in the same queue, taking always the one with the lowest key, independently of which search was being considered. Both methods work correctly [Goldberg and Harrelson, 2004].

The termination condition for bidirectional search is slightly different than Dijkstra's. Let μ be the minimum distance between s and t found so far. At the beginning, $\mu = \infty$. When an edge (u, v) is relaxed in the forward search, with node v previously settled in the backward search, then a path from s to t has been found, with a known distance. If this distance is less than μ, then μ is updated and v belongs to the shortest path found so far. The same occurs when a node u already settled by the forward search is reached by the backward search. Algorithm finishes when both searches settle the same node. At that point, the shortest path can be retrieved by traversing the predecessors (in both directions) of the node that triggered the last update of μ.

Comparing with Dijkstra's algorithm, this strategy is within a factor of two of the original algorithm in terms of efficiency (see [Goldberg and Harrelson, 2004] for details).

3.2.1   CASE STUDY

Figure 3.4 represents the bidirectional case of Dijkstra's algorithm. Light gray elements represent the forward search, whereas dark gray ones represent the backward one. Square-bounded numbers beside each node represent the Dijkstra's rank for the combined search.

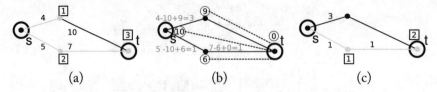

Figure 3.5: Settled nodes reordering from Dijkstra's to A*.

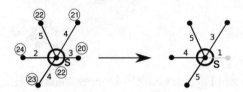

Figure 3.6: Starting step of A* algorithm.

3.3   GOAL-DIRECTED A* SEARCH

This approach [Hart et al., 1968] uses heuristics that try to avoid nodes not involved in the shortest path. Heuristics use domain information in order to look ahead in the decision process, trying to improve query times. However, if they are applied alone, their use does not always lead to shortest paths. As we will see, an appropriate heuristic combined with Dijkstra's algorithm ensures correct results.

In A* search, a heuristic (or potential) function $h'(v)$ returns a value that represents a lower-bound estimation of the remaining distance from that node v to the target one t. The query phase is as a normal Dijkstra's algorithm, where edge weights $w(u, v)$ are substituted with $w'(u, v) \leftarrow w(u, v) - h'(u) + h'(v)$. This new value represents the additional cost with respect to the heuristic estimation from the source, $h'(s)$, in case edge (u, v) is taken. The closer the lower bound value to the real one, the better the obtained results.

Dijkstra's algorithm uses distances from the source to reached nodes as the only criteria to decide which node will be selected next. Starting with the situation depicted in Figure 3.5 (a), Dijkstra's algorithm will choose the upper branch, while the lower branch belongs to the shortest path. On the other hand, A* estimates a lower bound distance from all reached nodes to the target (circle-bounded numbers are the Euclidean distances in Figure 3.5 (b)). With such an estimation, the transformation of edge weights $w'(u, v) \leftarrow w(u, v) - h'(u) + h'(v)$ leads Dijkstra's algorithm to a correct decision faster (see Figure 3.5 (c)). For route planning case in road maps, the Euclidean distance is a lower bound and therefore a valid estimator of $h(x)$. However, the search space is not always reduced enough to palliate the heuristic cost insertion (see experimental results in [Goldberg and Harrelson, 2004, Gutman, 2004]).

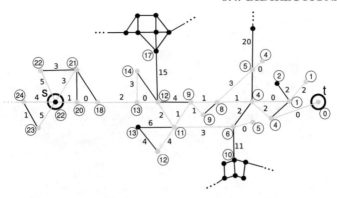

Figure 3.7: Settled nodes by A* algorithm. Weights are already transformed and circle-bounded numbers represent the Euclidean distance from each node to the target.

3.3.1   CASE STUDY

We will now show how A* works in our case study (see Figure 3.6). The algorithm starts from the source node, calculating, for every reached node r, an estimation $h'(r)$, shown in circle-bounded numbers. Then it transforms the edge weights (left) into weights that represent the additional cost in case the corresponding edge is taken (right). After that, the algorithm selects the edge with lower weight, highlighted in light gray, and repeats the process. Figure 3.7 represents in light gray all settled nodes of A* algorithm query using Euclidean distances as estimator.

3.4   BIDIRECTIONAL A* SEARCH

This approach uses a bidirectional search with two potential (heuristic) functions, one for forward search h'_s, and other for backward search h'_t. To use the termination condition that stops when both searches meet at the same node, proper potential functions should be selected. This issue leads to two different set of approaches: those that impose new termination conditions, called Symmetric approaches [Pohl, 1969], and those that use consistent potential functions, called Consistent approaches (see [Goldberg and Harrelson, 2005] for details).

3.4.1   CASE STUDY

Figure 3.8 shows the query phase applied to our study case using the symmetric approach. Circle-bounded numbers with right arrows represent the potential value of each node obtained during the forward search, whereas the ones with left arrows are those obtained during the backward search. Edge weights are already modified in the figure. Note that the algorithm does not stop when both searches settle the same node, but it continues exploring adjacent nodes (dashed edges in the figure), until every node in the queue will have their key $k(v) \geq \mu$, with $\mu = 10$ in our

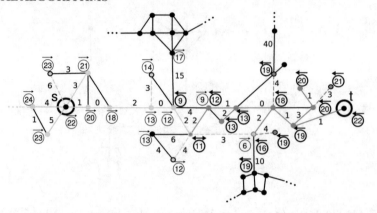

Figure 3.8: Settled nodes by a bidirectional A* query using symmetric approach.

case. On the other hand, if a consistent approach is used, the search will stop when both searches settle the same node.

CHAPTER 4

Hierarchical Preprocessing-Dependent Approaches

The hierarchical methods aim to discover a hierarchy in the graph that, for road networks, usually corresponds with their hierarchical nature. A graph hierarchy is a division of the graph nodes into levels. To define it, some precomputation is needed. There are different precomputations that can be done in the graph depending on the hierarchical algorithm chosen. The precomputation generates additional information that is attached to the graph, and it will be used in order to speed up the query time.

4.1 HIGHWAY HIERARCHIES

An intuitive way to reach a distant destination in a road network is to take the straightest highway roads from source to goal, thus implicitly using a hierarchy based on the road nature. The Highway Hierarchies (HH) approach [Sanders and Schultes, 2005, 2006a, 2012] follows a similar idea, classifying roads (edges) of the road network (graph) into levels of a hierarchy. The hierarchy of HH is created in a preprocessing phase. Although this method is no longer competitive, its fundamentals are integrated in other modern methods, such as Contraction Hierarchies or Transit Node Routing, that will be described later.

    The method relies on a tuning parameter H. Each node $s \in V$ has an ordered list of nodes with cardinality H that is called *neighborhood* of s, $N_H(s)$. These are the H closest nodes to s in the Dijkstra's rank Let w be the H-th closest node, and $d_H(s) = d(s, w)$ a distance function. The neighborhood can be defined as follows: $N_H(s) \leftarrow \{v \in V | d(s, v) \leq d_H(s)\}$. For directed graphs, an analogous concept of the neighborhood in the reverse direction is needed. Then, the "highway" concept is not related to the nature of the road, but to the connection between two neighborhoods.

4.1.1 PREPROCESSING: BUILDING THE HIGHWAY NETWORK

The duration of the preprocessing phase depends on the number of desired highway levels to be obtained. The built of a new level $G_i + 1$ involves two main processes: the first one decides which edges will belong to the new level, while the second one collapses the obtained graph, substituting paths with shortcuts.

1. *Searching for candidates.* The first process has two different phases. The first one is the construction of a *partial shortest-path tree* called B, that holds the candidates that will be evaluated. The second phase returns the edges that will belong to the following level, a process called *backward evaluation* of B. The algorithmic details to efficiently perform both phases are described in [Schultes, 2008]. The process starts from a source node s_0 and is repeated for the remaining nodes.

   (a) *Construction of B.* The objective is to create a shortest-path tree, called B, from source node s_0. Adjacent nodes to s_0 are called s_1 nodes. A "tagger" Dijkstra search is performed from source node, building the shortest-path tree while classifying the nodes into *active* or *passive*. The tagging process occurs when the nodes are being inserted in the queue. Node s_0 is tagged as active. A node v is tagged as passive if its parent p is passive. If p is active, v is tagged as active, unless the cardinality of the intersection of the neighborhoods of its ancestor node s_1 and p is less or equal than 1. The algorithm finishes when all nodes in the queue are passive. These nodes will not be added to the shortest-path tree. The leaves of the partial shortest-path tree B are called t_0 nodes.

   (b) *Backward evaluation of B.* Those edges (u, v) that belong to a path $P = \langle s_0, ..., u, v, ..., t_0 \rangle$ in the partial shortest-path tree B, with $u \notin N_H(t_0)$ and $v \notin N_H(s_0)$ will be considered as *highways*. These edges will be added to the resulting graph $G_1 = (V_1, E_1)$ that constitutes the following level of the hierarchy, together with their associated nodes.

2. *Contraction of Highway Network.* The contraction process is only needed when the current level will be used to obtain a higher one. This process modifies G_{i+1}, adding edges called *shortcuts* that allow us to jump from one neighborhood to another, avoiding all the intermediate nodes. In [Sanders and Schultes, 2005], this process was performed by seeking degree-one nodes, degree-two nodes, and attached trees, shortcutting them if it proceeds and extracting them from the next-level network. In [Sanders and Schultes, 2006a], the authors change the criterion to create shortcuts to a simpler method. Given a *contraction parameter* c, they bypass those nodes if the number of needed shortcuts is lower or equal than $c \cdot (deg_{in}(u) + deg_{out}(u))$.

The result of this process is the *contracted highway network* G'_{i+1}, also called *core*, which can be used as input for the next iteration of the construction procedure in order to obtain the following level of the highway hierarchy.

4.1.2 QUERY

The output of the preprocessing phase is a multilevel graph that will be used by the search algorithm to find shortest paths. The query algorithm is a modification of the bidirectional version of Dijkstra's algorithm. The algorithm proposed uses two priority queues, one for the forward search and another for the backward one. Every node in the queue will carry the following information: The tentative distance from source/target; the level of the search (defined as the level of the last edge traversed to reach that node); and the remaining distance to neighborhood limit.

The algorithm proceeds as follows. First, starting from both the source and destination nodes, it performs a Dijkstra search in the subgraph of the H-closest nodes (neighborhoods). When settling a node u, if there exists an edge $e = (u, v)$ that leaves the current neighborhood (that is, when u's remaining distance to neighborhood limit is lower than the edge weight), node u is called an *entrance point*. In this case, the current level l will be increased until the size of the new neighborhood of u in G_l is higher than the weight of e. Then, if the level of the edge e is lower than the reached level l, e is not relaxed. Otherwise, e is relaxed and node v will be inserted in the queue with (a) the new level and (b) the remaining distance to u's neighborhood limit in level l. If entrance point u does not belong to the core of level l, $u \notin V_l'$, its neighborhood size is considered as infinite. In this case, the search continues in the current level until it reaches a core node c. Then the following nodes reached from c will adopt the remaining distance to c's neighborhood limit and the search continues proceeding as described above. In order to speed up queries, when a core node is settled, edges $e = (u, v)$ with $u \in V_l'$ and $v \in V_l \wedge v \notin V_l'$ are not relaxed.

Note that the search can find different entrance points that would expand the search to different exploration levels, leading to a priority queue with nodes with different levels. The queues are sorted by the tentative distance as usual. In case of tie, the node with higher level will have higher priority. For ties in the level value, the one with less remaining distance to neighborhood limit will have higher priority. Note that other tie-breaking rules also lead to a correct algorithm.

The process does not stop when both searches meet, but continues until all tentative paths have more cost than the best solution found so far.

The main advantage of the algorithm is that, when s and t are far away, the search is forced to use highway edges. According to the authors of the method, this leads to query times in order of milliseconds. Note that, once the query has finished, the algorithm should construct the complete shortest path, a non-trivial task [Sanders and Schultes, 2006a].

4.1.3 OPTIMIZATIONS

There are some techniques, described in Sanders and Schultes [2012], can be applied in order to improve the HH performance:

1. *Rearranging nodes* with respect to their core level leads to improving the locality property in higher levels, that in turn leads to smaller query times.

2. *Precomputing distance tables* for the highest level allows the algorithm to avoid the search in this level, by looking up these tables instead of computing the distances (see Sect. 4.2).

3. *Abort criteria.* In Sanders and Schultes [2005], more sophisticated termination criteria that reduce the explored search space are described. However, the simpler criteria described above can be efficiently evaluated and, therefore, yields better query times despite the larger space exploration.

4.1.4 CASE STUDY

In order to show in detail how the HH preprocessing phase works, we will not consider the dark area close to the source node. As we will see later, this simplification will not lead to noticeable differences, but simplifies the explanation.

Preprocessing   Phase 1(a) of preprocessing ("construction of partial shortest-path tree B") is illustrated in Figure 4.1 (a). The parameter H used in the example is four. Therefore, each neighborhood will have five nodes that are represented by big circles. The figure reflects the result of the phase. Let us focus on the light gray rounded node. This is the first node outside the neighborhood of node s_1, also depicted in light gray. The neighborhood of the highlighted node has two nodes in common with the neighborhood of s_1, therefore it is still active. After that step, the neighborhood of the following nodes to be analyzed do not have any elements in common with the s_1 neighborhood. Therefore, these nodes become passive and the "candidate search" is stopped because there are no more active nodes in the queue.

The resulting edges from phase 1(a) will be examined in phase 1(b) to consider them as highways or not. From all candidate edges of phase 1(a), only those that fulfill the condition are highways. Figure 4.1 (b) shows the partial highway network (G_1) corresponding to Figure 4.1 (a), generated from s_0. All edges connecting u nodes with adjacent v nodes in the figure are the last ones that fulfill the condition, because nodes v and beyond will be inside the neighborhood of t_0.

Phases 1(a) and 1(b) that we have shown for s_0 are repeated for all remaining nodes. After the creation of each partial shortest-path tree B, new edges that had not appeared yet in G_1 are added. Figure 4.2 shows the resulting level-one graph G_1 from process 1 ("searching for candidates") of HH precomputation.

The *contraction process* of G_1 is illustrated in Figure 4.3 (a) and (b). The bypassability criterion was $\#shortcuts < deg_{in}(u) + deg_{out}(u)$. Figure 4.3 (b) shows the resulting graph G_1', that serves as input graph to construct G_2, through process 1. Figure 4.3 (c) and (d) show the resulting graph G_2 and its contracted graph G_2', respectively. If the user would like to obtain a G_3 graph, G_2' should be processed as well.

Query   Figure 4.4 represents three different levels. The local searches start in G_0 (Figure 4.4, bottom), with forward search in light gray and backward in dark gray, traversing the corresponding neighborhoods. In the example, each search, after settling its neighborhood nodes, switches to

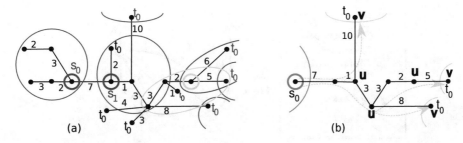

Figure 4.1: HH example, (a) phase 1a: Candidate search computing from a node s_0, and (b) phase 1b: Backward evaluation, obtaining the part of the highway network G_1 that starts from s_0. Highway edges are highlighted with dashed arrows.

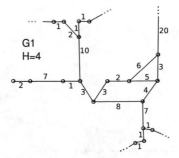

Figure 4.2: HH example after process 1: The complete highway network G_1.

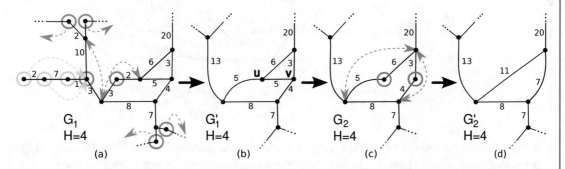

Figure 4.3: HH example: Contraction process of G_i, leading to the contracted graph G'_i.

the following level G_1 (Figure 4.4, middle) through its entrance point (dashed circles). Dotted lines in Figure 4.4 represent these level changes.

The entrance point of the forward search is not a core node of G_1, so it has no neighborhood limit. The search continues until it reaches a core node. Regarding the backward search, the entrance point to level G_1 is a core node in that level (HH_7), so it continues the search in its core

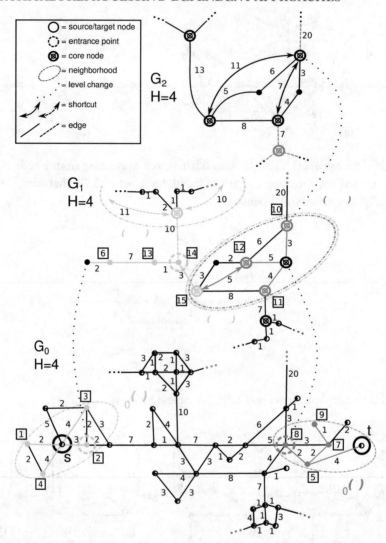

Figure 4.4: Settled nodes of an HH query, representing all traversed levels. Note that dotted lines are not edges, but level changes. Dashed edges and dashed shortcuts represent edges that are explored after both searches meet.

neighborhood $N_1(HH_7)$. When forward search settles core node HH_2, its neighborhood limits are reduced to its core neighborhood $N_1(HH_2)$. After that, backward search settles also HH_2.

Note that although both searches have met in level 1, they continue (dashed edges) until all remaining nodes in the queues have a higher tentative distance than the last minimum distance found, even at higher levels. While in [Schultes, 2005] more powerful termination criteria are

Table 4.1: HH example with Distance Tables between core nodes of level G_2

G_2'	G_2'	Distance	G_2'	G_2'	Distance
HH_1	HH_2	13	HH_2	HH_4	8
HH_1	HH_3	24	HH_2	HH_6	15
HH_1	HH_4	21	HH_3	HH_4	7
HH_1	HH_6	28	HH_3	HH_6	14
HH_2	HH_3	11	HH_4	HH_6	7

described, [Schultes, 2008] shows that the criteria used here are simpler and more efficient. In order to present a clean picture we have omitted the entrance-point mark of HH_3 and HH_4, and the level change in HH_4.

Considering now the original study case, including the removed dark area, the construction process generates few more highways, transforming the attached tree present in G_1 into a line that will be shortcutted in level two. The query will then enter in the dark area, settling as many nodes as the value of the neighborhood size per level.

4.2   OPTIMIZING HIERARCHICAL QUERIES WITH DISTANCE TABLES

In a hierarchical method, if the graph corresponding to the highest hierarchy level has a reduced size, it is not so expensive to compute and save all the distances between the nodes of this level. In this solution, proposed by [Sanders and Schultes, 2006a], when the algorithm reaches this level, both searches stop, and a few table look-ups are performed, achieving a slight speed-up compared with the pure hierarchical version. As an example, Table 4.1 shows the shortest-path distances between the nodes belonging to G_2' in our case study.

4.3   CONTRACTION HIERARCHIES

Contraction Hierarchies (CH) [Geisberger et al., 2008] use the node contraction idea of Highway Hierarchies to create a graph hierarchy.

The key idea of the method is to find a priority total order $rank : V \rightarrow 1..n \in \mathbb{N}$ for all the nodes of the graph based on a given criteria. In a weighted directed graph $G = (V, E)$ with n nodes, the method numbers the nodes from 1 to n according to the criteria used. Obtaining this order is the first part of the preprocessing phase. The second part (contraction process) removes nodes, keeping their order and adding shortcuts in the new resulting graph. The total priority order defined drives the query phase, leading the search to higher-priority nodes, thus reducing the search space.

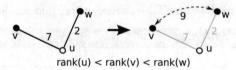

Figure 4.5: CH contraction process example.

4.3.1   PREPROCESSING: CONTRACTING THE NETWORK

The contraction process starts shrinking the graph, removing the node v with less priority. In order to preserve shortest paths on the new resulting graph $G' = (V', E')$, the process generates shortcuts between all nodes that are adjacent to v. The weight of each shortcut represents the sum of the weights of the two edges that the shortcut is replacing (see Fig. 4.5).

To decide if a shortcut should be added, a local search is performed from each $u : (u, v) \in E$ toward every node $x : (v, x) \in E$, checking the following conditions:

- The path $\langle u, v, x \rangle$ is the only shortest path from u to x. In another case, the insertion of the shortcut is prevented by another path P with weight $w(P) \leq w(\langle u, v, x \rangle)$, called *witness path*.

- An edge (u, x) does not exist. Otherwise, no shortcut is added, but the edge weight $w(u, x)$ is reduced to the shortest-path weight, $w(u, v) + w(v, x)$.

Note that we can stop the search from u to all x when distance $w(u, v) + max\{w(v, x) : (v, x) \in E' \setminus \{(v, u)\}\}$ is reached. However, this search could be notoriously time consuming, especially when very long edges (as ferry connections in transportation networks) are involved. The authors propose two heuristics to avoid these expensive situations. The first one is to limit the number of settled nodes, using a Dijkstra search that can be stopped after a certain number of nodes are settled. The second one is to limit the number of hops/edges from the starting node. In this way, the paths found by the search are limited to a certain number of edges, called *hop limit*. Although these limitations do not invalidate the algorithm, in some cases, they derive in adding more shortcuts than is really needed, and therefore it leads to a more dense graph with a higher contraction time (see [Geisberger, 2008] for more details).

4.3.2   PREPROCESSING: CHOOSING THE BEST ORDERING

Any possible order leads to a correct development of the contraction process. However, the difference between them in terms of preprocessing and query performance can be huge. In order to estimate how high should be the priority of a given node, we need to compute a linear combination of several properties:

Edge difference Considered by the authors as the most important priority term [Geisberger, 2008], this property is the difference between the number of edges of the current graph and the number of edges of the graph with the selected node contracted. A node is more suggestive to be contracted if its contraction supposes a big reduction of edges.

Uniformity The contraction process should choose nodes that are uniformly distributed in the graph, rather than select them always from small regions. Two independent successful heuristics are: *Deleted neighbors*, taking into account the number of already contracted neighbors; and *Voronoi-Region parameter*, using the square root of the Voronoi region size of the node as a priority term.

There are other optional properties that can be used to refine the priority of a node in order to obtain better quality hierarchies, or to reduce some high-consuming times of the contraction phase: *Cost of contraction*, adding the temporal cost of a node contraction to its priority linear combination; *Cost of queries*, adding to the linear combination a property that eases the query phase, by reducing the depth of the shortest-paths trees raised in this phase; *Global measures*, using betweenness or reach concepts to find the less important nodes. See [Geisberger, 2008] for details.

4.3.3   PREPROCESSING: BLENDING THE ORDERING PROCESS AND THE NETWORK CONTRACTION

Note that if node ordering is based in properties related to the contraction process, such as edge difference, performing the contraction process changes the graph nature. Therefore, the properties of the remaining nodes and their order will also be affected. To handle this *updating problem*, one of these techniques should be used:

1. *Lazy update* involves the updating of the lowest node v in the priority queue before its contraction. If its new priority is still lower than the following node u, v is contracted. In another case, v is reinserted in the queue and the process is repeated on u until a consistent minimum is found.

2. *Neighbors recomputing* updates the contracted-node neighbors priorities.

3. *Periodically rebuilding* updates all nodes and rebuilds the queue.

4.3.4   QUERY

The query algorithm is a variation of a Dijkstra bidirectional shortest-path search. It alternates between forward and backward search. The forward search explores the *upward graph* $G_\uparrow = (V, E_\uparrow)$, with $E_\uparrow = \{(u, v) \in E : rank(u) < rank(v)\}$. The backward search explores the *downward graph* $G_\downarrow = (V, E_\downarrow)$, with $E_\downarrow = \{(u, v) \in E : rank(u) > rank(v)\}$.

When a node previously settled in one direction is settled by the other direction, a new shortest-path candidate is obtained. The algorithm finishes in one direction if the weight of the first element in the queue is at least as large as the weight of the best shortest path found so far.

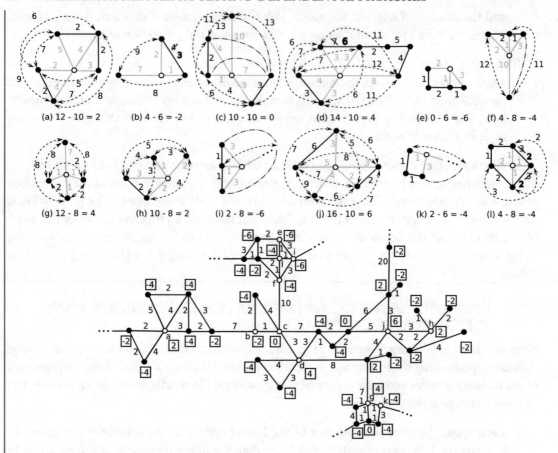

Figure 4.6: CH initial simulations (top) and heuristic results (bottom).

The search space can be pruned using the *stall-on-demand* technique [Schultes and Sanders, 2007]. Let $\bar{d}(v)$ be the accumulated weight associated to node v when it is going to be settled. In a forward search, when v is going to be settled, the algorithm checks if $\bar{d}(v)$ is the shortest distance from source s, $d(s, v)$, using the edges (x, v) of $G_\downarrow$ with $rank(x) > rank(v)$ as follows: If $\bar{d}(x) + w(x, v) < \bar{d}(v)$ then there is a path to v shorter than the one found so far. Therefore, the search through v can be pruned (*stalled*). Although the stalled nodes are settled with the lower distance, the search does not relax their incident edges (see [Geisberger et al., 2008] for more details).

4.3.5 CASE STUDY

Node ordering and contraction   In order to find a node ordering for our case study, we will use the *edge difference* heuristic. Computing this term for all nodes implies simulating a contraction

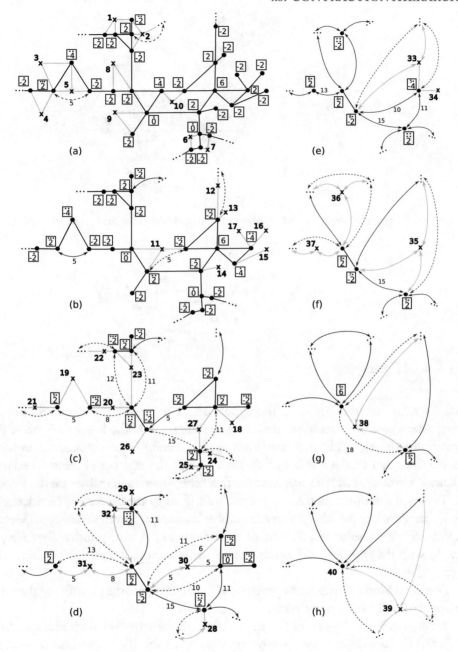

Figure 4.7: CH contraction steps.

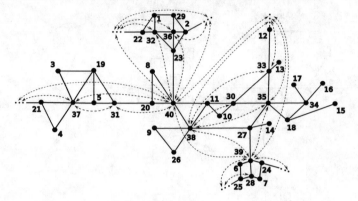

Figure 4.8: CH preprocessing output graph, with the node ordering and shortcuts added.

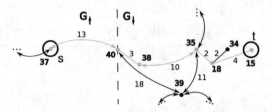

Figure 4.9: CH query phase.

process for each one (see Fig. 4.6). The upper part of the figure represents the simulation of the contraction process of some selected nodes (white nodes). The lower part shows the graph annotated with the edge-difference heuristic result for all nodes (square-bounded numbers). These values are the result of the number of shortcuts minus the number of removed edges. In the simulations, arrow-dashed lines represent the first term, whereas gray lines are the second term.

To obtain a uniform contraction process we will apply the *deleted neighbors* technique (see Fig. 4.7). The dots into the edge-difference number boxes represent the number of deleted neighbors. We will use this value as a tie-breaker for nodes with the same priority. Therefore, between two nodes with the same edge difference, the one with lower priority will be the one with fewer deleted neighbors.

Finally, to handle the updating problem, we will recompute the priorities of the contracted-node neighbors after every contraction.

Figures 4.6 to 4.8 shows how the preprocessing phase is applied to our example. In Fig. 4.6, (b), (d), and (l) have edges whose weights have changed; (c), (f), (i), and (k) have examples of shortcuts not added (point-line arcs) because the path $\langle u, v, x \rangle$ is not the shortest one; (d) and (e), among others, show the existence of long witness paths, thus preventing the addition of shortcuts. Figure 4.7 shows some snapshots of the contraction process while obtaining the node ordering. Figure 4.8 shows the final node ordering and the output graph of the preprocessing phase.

Query   Figure 4.9 shows the query phase. The forward search traverses the upward graph $G_\uparrow$, alternating with the backward search that traverses the downward graph $G_\downarrow$. Both meet in the node with rank 40. At that moment a candidate for the shortest path has been found with weight 32. Note that, in our example, this is actually the real shortest path. However, the search continues in the dark areas until the first element in each search queue has a weight greater than the weight of the candidate we have found.

4.4   TRANSIT-NODE ROUTING

In queries where the source is "far enough" from the target, it is not uncommon that many of them traverse the same nodes. These nodes are called *transit nodes*. Combining this concept with Distance Tables, we obtain the Transit-Node Routing (TNR) approach [Bast et al., 2007].

4.4.1   TRANSIT-NODE SET SELECTION

There are different methods to select which nodes are transit nodes, i.e., separators [Delling et al., 2006a], border nodes of a graph partition [Bast et al., 2006, 2007], and nodes that belong to a high level in a hierarchy created by a hierarchical method, as Highway Hierarchies [Bast et al., 2007, Sanders and Schultes, 2006b] or Contraction Hierarchies [Geisberger, 2008, Geisberger et al., 2011]. The following descriptions are focused in the hierarchical based transit-nodes method because they return better performance in road networks compared with the others. The transit-node selection done with Contraction Hierarchies has even better performance than using Highway Hierarchies, reducing preprocessing and query times and space consumption.

These "high-level nodes" have the property that they are going to be used many times for a set of long queries. Therefore, it is possible to define the transit-node set T taking the core nodes of one of the highest levels.

In order to know if a particular source-target query is a long one, locality filters are defined associated with the transit node set, L for T. The locality filters return true if the source and the target belong to a local area. This means that they are not far away enough to use T. Otherwise, if it returns false, T can be used to compute the shortest-path distance.

As long as the long queries are going to be performed quickly with few table lookups, now the problem is the local queries. In order to palliate it, more layers of transit nodes can be defined. Since K is the hierarchical level used to define the transit node set, that is level K for the first layer T_1, the authors in [Bast et al., 2007] have used level $\lceil K/2 \rceil$ for second layer T_2 and level $\lceil K/4 \rceil$ for third layer T_3. Note that $T_1 \subset T_2 \subset T_3$.

Once the transit nodes have been chosen, the *access nodes* for each node s are defined as the transit nodes whose paths from s do not traverse other transit nodes. These access nodes will act as access points from s to the transit network.

4.4.2 PREPROCESSING

The preprocessing phase consists in computing (a) a table for each node s with the shortest-path distances between s and its access nodes; (b) a table with the shortest-path distances between each pair of transit nodes; and (c) the locality filter.

There are two methods to compute the first table: *forward* and *backward approaches*. The *forward approach* performs a Dijkstra search from each node u that stops when all nodes in the priority queue are *covered* by transit nodes. A node is covered if it has a transit node as an ancestor. The transit nodes found that are not covered compound the access node set of u, $A(u)$. It is possible to improve the efficiency of this method by applying two tricks (see [Bast et al., 2007] for details).

The *backward approach* performs a backward Dijkstra search from each transit node v that also stops when all nodes in the priority queue are covered by transit nodes. Then, v is stored as an access node for every not-covered node u, together with its distance. During this process we can create edges (v, x) for a transit graph $G(T)$ in which weights hold the shortest-path distances from the transit node v to other transit nodes x. After performing this process for all transit nodes, the transit graph is complete. Note that this graph $G(T)$ is the same as the graph of the selected K hierarchical level.

The computation of the transit-node table distances can be performed using any standard all-pairs shortest-path algorithm on the transit graph. The information obtained when computing distances of transit-node layer $i + 1$ is very useful for computing an effective locality filter of level i. This method calculates for every node $u \in T_{i+1}$ a circle that contains all meeting points (of bidirectional searches) with the nodes $v \in T_{i+1}$ that have a distance entry $d(u, v)$ in the layer $(i + 1)$ table. The radius of this circle is denoted as $r^i(u)$. Now for any node x we can compute a circle radius $r^i(x) \leftarrow \max\{d_E(u, x) + r^i(u) : u \in A_{i+1}(x)\}$ where $d_E(u, x)$ represents the Euclidean distance between u and x.

The authors in [Bast et al., 2007] have used for the hierarchical transit-node selection case (a) the forward approach to compute the access-node sets, with highway searches instead of Dijkstra searches; (b) a many-to-many routing algorithm [Knopp et al., 2007] to compute the transit-node distance layers; and (c) the previously described method to compute the locality filters.

4.4.3 QUERY

For a given query from source s to a destination t, the algorithm first retrieves their corresponding access-node sets A_s and A_t. With this information, only a few look-ups in the distance tables return the distance $d(s, t) = min\{d(s, a_s) + d(a_s, a_t) + d(a_t, t) : a_s \in A_s, a_t \in A_t\}$. Afterward, it checks if they belong to a local area using the locality filter associated to the first layer of transit nodes. If it returns false, s and t are far enough, and the algorithm finishes. Otherwise, s and t are in a local area, so the process is repeated for the following layer. If the last locality filter returns true, the algorithm performs a bidirectional highway hierarchy search that stops when the radius returned by the forward and backward search surpasses the lowest value computed in previous layers. For algorithms not based in hierarchical transit nodes, any shortest-path algorithm can be performed to compute the distance inside the local area.

The precomputation of all the information required for every node and the obtention of many transit node layers with their distances lead to very good query times, in order of microseconds (see results of [Bast et al., 2007]). Unfortunately, this efficiency is at the cost of longer preprocessing times and greater space consumption. The transit-node routing method described above, that stores all this information, is called the *generous variant*. A more economic variant is described below.

4.4.4   ECONOMIC VARIANT

There are some guidelines that reduce the space consumption at the cost of slower query times. Their implementation leads to a different transit-node routing version called the *economic variant*.

This variant computes fewer layers than the generous variant. Moreover, the level chosen for the first layer of transit nodes is the highest level of the hierarchy. In this way, it decreases the size of stored tables by reducing the number of transit nodes and the distances that are going to be stored. Besides this, for the first layer, the locality filter and the access-points tables for each node and the corresponding distances are not stored. Instead, they are computed on-the-fly in the query phase using the information of the second layer.

In the experiments carried out in [Bast et al., 2007], this economic variant reduces the space used by the generous one to a half.

4.4.5   CASE STUDY

Figure 4.10 shows our study case. Gray rounded nodes are the selected transit nodes. Light gray and dark gray rounded nodes are, respectively, the access points of source and target nodes. Dashed lines represent distances stored in tables, both between each node and its access nodes (light gray and dark gray dashed lines) and between all transit nodes (black dashed lines). Tables 4.2 and 4.3 show the corresponding values that are stored.

4.5   HUB-BASED LABELING ALGORITHM

The Distance Labeling technique labels each node of the graph with distance information regarding the shortest paths between nodes. Although this labeling leads to very fast queries, its drawback is the cost of the labeling process in terms of time and space, making this technique impractical. The Hub-based labeling algorithm [Abraham et al., 2011] takes advantage of Contraction Hierarchies (see Sect. 4.3) and other methods to reduce the cost of the Distance Labeling approach for point-to-point queries, leading to a particularly fast method.

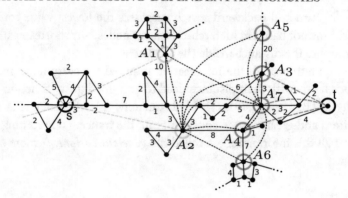

Figure 4.10: Stored distances in TNR algorithm approach from source and target node to their access points. As an example, distances from A_2 and A_7 to the rest of transit nodes are also shown.

Table 4.2: TNR example: Distance Table between transit nodes

TN	TN	Distance	TN	TN	Distance	TN	TN	Distance
A_1	A_2	13	A_2	A_4	8	A_3	A_7	3
A_1	A_3	24	A_2	A_5	31	A_4	A_5	27
A_1	A_4	21	A_2	A_6	15	A_4	A_6	7
A_1	A_5	44	A_2	A_7	10	A_4	A_7	4
A_1	A_6	28	A_3	A_4	7	A_5	A_6	34
A_1	A_7	23	A_3	A_5	20	A_5	A_7	23
A_2	A_3	11	A_3	A_6	14	A_6	A_7	11

Table 4.3: TNR example: Distances from source and target to their access nodes

Node	Access point	Distance
s	A_1	23
s	A_2	16
t	A_3	9
t	A_4	10
t	A_7	6

4.5.1   DISTANCE LABELING

The distance labeling scheme for a graph $G = (V, E)$ [Gavoille et al., 2001] has two elements: A *node-labeling function* $L()$ that assigns to each node $v \in V$ a label $L(v)$, with distance information; and a *distance-decoder function* $f()$ that obtains the shortest-path distance of a point-to-point query using the labels of both the source $L(s)$ and target $L(t)$ nodes. The Hub-based labeling algorithm uses the CH search as the node-labeling function (storing the settled nodes and their distances), and the intersection of the sets as the distance-decoder function.

4.5.2 PREPROCESSING: COMPUTING LABELS

One of the main problems of distance labeling is its huge storage requirements. The use of CH searches stretches the label size of a node v to those nodes with a higher rank than v. The *forward label* $L_f(v)$ will contain all nodes settled in a forward CH search with v as source node, together with their distances. The *backward label* $L_b(v)$ is defined analogously, using a backward CH search. Note that forward and backward labels of a node v can differ in directed graphs, while are equal in undirected ones.

A forward label $L_f(v)$ is composed by three elements: an integer N_v, representing the number of nodes in the label; an array I_v, with the IDs in ascendant order of all nodes in the label; and an array D_v, with the distances from v to each node in the label, that is, $D_v[i] = d(v, I_v[i])$. Backward labels $L_b(v)$, are defined analogously, using the distances from all nodes to v, that is, $D_v[i] = d(I_v[i], v)$.

The label size can be reduced further using the *stall-on-demand* technique [Geisberger et al., 2008] that prunes the distances that are greater than the shortest one. The authors [Abraham et al., 2011] use more techniques to reduce the label average size even more.

4.5.3 QUERY

Given a query from s to t, the algorithm intersects the sets of nodes stored in the respective labels $L_f(s)$ and $L_b(t)$. Each $v \in L_f(s) \cap L_b(t)$ indicates that exists a path from s through v to t. Since i and j are the indexes of v in labels $L_f(s)$ and $L_b(t)$, respectively, the algorithm takes the one that minimizes $D_s[I_s[i]] + D_t[I_t[j]]$, that is, the shortest-path distance from s to t. Since both arrays I_s and I_t are sorted, this process can be done traversing both arrays in one sweep.

This process returns the shortest-path distances, together with the intermediate nodes. To recover the whole shortest path the authors use the technique exposed in [Bast et al., 2007].

Recently, the Hierarchical Hub Labeling method [Abraham et al., 2012] proposes new node-set ordering algorithms, more efficient labeling processes, and pruning criteria. This method reduces the average label size even more.

4.5.4 CASE STUDY

Preprocessing labels Figure 4.11 shows the CH search from some example nodes and their corresponding labels, with bold numbers representing the CH node ranks as shown in Fig. 4.8. Note that, because the graph is undirected, all labels $L_f(v)$ and $L_b(v)$ are the same for each node v. The search performed in the query phase of CH (see Fig. 4.9) also allows us to obtain labels $L_f(s)$ and $L_b(t)$, being $N_s = 1$, $I_s = [40]$, $D_s = [13]$, and $N_t = 6$, $I_t = [18, 34, 35, 38, 39, 40]$, $D_s = [4, 6, 6, 16, 17, 19]$ respectively.

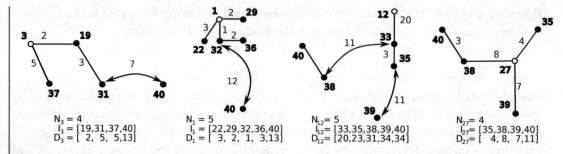

Figure 4.11: HL computing labels.

Query   The query phase just intersects both sets, $L_f(s)$ and $L_b(t)$, and obtains that the shortest path is through node v, with $rank(v) = 40$. Therefore, the distance of the shortest path from s to t has a length $13 + 19 = 32$.

4.6   OTHER HIERARCHICAL PREPROCESSING METHODS

There are other techniques that fall into this category. Highway Node Routing (HNR) [Schultes and Sanders, 2007], that is based on Highway Hierarchies with different preprocessing criteria, is the ancestor of Contraction Hierarchies. A novel variant of TNR [Arz et al., 2013] combines the methods used in the CH approach with the DT technique. There, the transit-node set is made by selecting the upper T nodes of the contraction order, and then, local queries are solved using CH, instead of HH.

CHAPTER 5

Non-Hierarchical Preprocessing-Dependent Approaches

The non-hierarchical preprocessing methods aim to avoid settling unnecessary nodes using information obtained in a preprocessing phase that does not follow a hierarchical structure. The nature of these approaches is diverse, extracting different sets of data during precomputation. The following subsections will describe the approaches that fall into this category.

5.1 LANDMARK A*

Landmark A*, or ALT [Goldberg and Harrelson, 2005], improves the search of A*, leading to better lower bounds. This method combines A* with the use of landmarks and the triangle inequality. *Landmarks* are an arbitrary subset of nodes whose distances from/to all nodes are known after the preprocessing phase. Regarding the triangle inequality, it says that the sum of the lengths of any two sides must be greater than or equal to the length of the remaining side. Knowing the shortest path from/to a landmark L to/from two different nodes v and w, the distance $d(v, w)$ can be estimated using a slightly different version of the triangle inequality. Nodes L, v, and w form a triangle (see Figure 5.1). Knowing $d(L, v)$ and $d(L, w)$, the following inequalities are true and can be used to guide the search:

$$d(v, w) \geq d(L, w) - d(L, v) \tag{5.1}$$
$$d(v, w) \geq d(v, L) - d(w, L) \tag{5.2}$$

Note that, if the weight of the path from v to w were smaller ($d(v, w) < d(L, w) - d(L, v)$), then the shortest path from landmark L to node w would be the subpath from L to v plus the subpath from v to w, ($d(L, v) + d(v, w) < d(L, w)$), that would be shorter than that obtained in the preprocessing phase.

5.1.1 LANDMARK SELECTION

The first step is to compute the set of landmarks. Although the choice of landmarks could be done randomly, a proper selection will lead to better lower bounds. This is the only step where domain

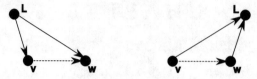

Figure 5.1: Different scenarios in the estimation of $d(v, w)$ using the triangle inequality.

knowledge is taken into account. Algorithms for this choice range from random choice to *farthest landmark selection* (that iteratively selects the farthest node to the current landmark set), to *planar landmark selection* (that tends to choose equidistant landmarks). See [Goldberg and Harrelson, 2004] for details.

But the best heuristics are *avoid* and *maxCover* [Goldberg et al., 2007]. Briefly described, the *avoid landmark selection* method improves an existing landmark set S adding new ones in the regions of the graph that are not "well-covered." The *maxCover landmark selection* uses the *avoid* method to generate a set of landmarks. Afterward it evaluates its quality and tries to improve it, replacing the useless landmarks for better ones.

5.1.2 PREPROCESSING: COMPUTING LANDMARK DISTANCES

After the landmark choice, the preprocessing phase is very simple. It just computes the bare Dijkstra's algorithm, in forward and backward searches, to obtain the distances from/to landmarks to/from nodes.

5.1.3 QUERY

With the distance from/to landmarks, for each node v and for each landmark $l \in L$ two estimations of the distance to the target, $d(v, t)$ can be computed, using the inequalities (5.1) and (5.2). The best estimation is the highest one. Thus, the potential function of a node v is $\max_{l \in L}(\max([d(l, t) - d(l, v)], [d(v, l) - d(t, l)]))$. The query phase is identical to the one used in A* heuristic, with the difference that now better values for the estimations (higher lower bounds) are used instead of using the Euclidean distance as heuristic.

Regarding relative performance, [Goldberg and Harrelson, 2004] reported that the use of a consistent bidirectional ALT algorithm with the average potential function leads to the best efficiency.

5.1.4 CASE STUDY

Landmarks selection and preprocessing   Regarding our example (Figure 5.2 (a)), spiral nodes are the landmarks chosen. In preprocessing phase, all distances from/to landmarks to/from the rest of nodes are computed. The study case is an undirected graph, so both distances are the same.

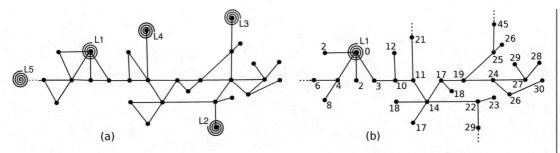

Figure 5.2: (a) Selected landmarks. (b) Dijkstra's shortest-path tree from L_1 with the respective distances.

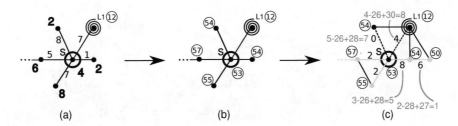

Figure 5.3: Edge weight transformation in query first steps.

Figure 5.2 (b) shows the shortest-path tree from landmark L_1, with the distances to the rest of nodes.

Query   Figure 5.3 represents the first steps of the query phase, and the change of edge weights when the heuristic information is applied. Figure 5.3 (a) shows the distance from landmark L_1 to the nodes (bold numbers). A lower bound for the distance from source to target, $h'(s)$, can be obtained as $d(L_1, t)$ $d(L_1, s)$. Figure 5.3 (b) shows in circle-bounded numbers the lower bounds for the nodes reached from s, whereas Figure 5.3 (c) shows the graph with the transformed weights of the edges ($w'(u, v) \leftarrow w(u, v)$ $h'(u) + h'(v)$). The algorithm will first visit those nodes with zero transformed distance, settling nodes in the dark area until it finds a transformed weight higher than the ones shown in Figure 5.3 (c). If we only use L_1 landmark, the search in the dark area will settle all nodes inside the radius of the distance $d(L, t)$, till the estimations from the nodes become negative, indicating that the target cannot be found following these paths. Figure 5.4 shows the settled nodes and edges in light gray. Weights of edges with an associated transformed weight equal to zero are omitted. As can be seen, the query using just one landmark does not lead to a noticeable reduction in the number of nodes settled. The more landmarks we use, the more efficient the query will be. Besides this, note that L_1 is not very well situated with respect to s. Informally, landmarks should be "behind" the node being evaluated or "beyond" the target to give good estimations.

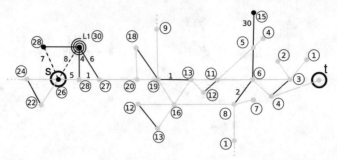

Figure 5.4: Settled nodes of the ALT algorithm with only one landmark.

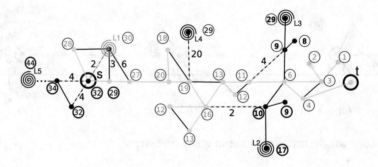

Figure 5.5: Settled nodes of the ALT algorithm using all landmarks.

Figure 5.5 shows how the addition of more landmarks improves the query phase. Dashed edges were avoided thanks to the estimation provided by the landmark closest to each node.

Note that the use of L_5 improves the potential of node s and some adjacent nodes because the situation of the landmark is behind them. Now, the search is not only capable of avoiding the two first black dashed edges and the entrance to the dark area, but also to lower the transformed weights that lead to the target. The addition of other landmarks, L_2, L_3, and L_4, serves to avoid their respective black dashed edges and to stop the entrance of the search in their respective dark areas.

5.2   EDGE FLAGS

Road traffic posts notify the drivers which path should be taken in an intersection to reach a particular destination. Following the same idea, the Edge Flags algorithm (also known as Arc Flags) [Hilger et al., 2009, Köhler et al., 2005, Lauther, 2004, Möhring et al., 2005] divides the graph in r regions so that each node v belongs only to one of them, and takes a specific route depending upon the region where the target node is located. To develop the idea, roads (edges) belonging to a shortest path that enters into the target region should be tagged. In this way the algorithm would just have to consider a small quantity of tagged edges during the search.

For each edge $e = (v, w)$ and each region i, there are two flags, $v\_flag_{e,i}$ and $w\_flag_{e,i}$. These flags indicate if the edge belongs to a shortest path from v or w using e into region i. Having m edges, there will be $m * r * 2$ flags for the whole graph. Thus, when the algorithm relaxes the edges of a settled node, it checks if they belong to a shortest path toward the target region. If one edge does not have the corresponding flag set, then it is pruned.

5.2.1   REGION DEFINITION

The division in regions is crucial for the performance of the algorithm. In [Lauther, 2004] two methods are proposed: *Grid-based region* (divides the map into squares of equal width) and *Square covering* (divides the map in overlapped rectangular regions). In [Hilger et al., 2009] the authors compare partition approaches: *Grid, quad-trees, kd-trees,* and *multi-way arc separators* (*METIS* [Karypis, 1995] and *PARTY* [PAR, 2006]). According to their results, the best methods are kd-trees (the *median* variant) and multi-way arc separators, but the former is easier to implement.

A better partitioning algorithms can be applied to perform the region definition as the recent PUNCH [Delling et al., 2011a] algorithm. The authors state that PUNCH gives the best partitions for the road maps at the cost of a slower preprocessing compared with the other partition algorithms.

5.2.2   PREPROCESSING: RAISING FLAGS

Computing all the flags of the graph is a time-consuming task due to the high number of shortest paths. The brute force approach is to compute all the possible shortest paths from every node in a region $n \in r_i$ to the rest of nodes using a Dijkstra backward search. This aims to find the edges that belong to the shortest paths from nodes outside of region r_i to those inside of it, obtaining the reversed shortest-path tree. A faster method, described in [Lauther, 2004], is based on the idea that only a subset of nodes allows the search to to enter into a region. Edges are divided into two categories: *Interface edges* that connect nodes from different regions; and *internal edges* that connect two nodes of the same region. Nodes adjacent to interface edges are *exported/boundary nodes*, and the rest are *interior nodes*.

To reach a region, it is necessary to traverse one of these boundary nodes. In this way, the preprocessing phase only builds reversed shortest-path trees from each boundary node. Taking advantage of the similarity of the resulting trees from close boundary nodes, the preprocessing step was improved with the so-called *pruned shortest-path trees* [Möhring et al., 2005]. The shortest-path tree of one of the boundary nodes v_1 represents an upper bound for the construction of the closest boundary nodes trees. Since v_2 is another boundary node, it is known that the nodes in the shortest-path tree will have a length equal to the distance in v_1's shortest-path tree plus the distance between v_2 and v_1, or even less, if v_1 is not in the middle of the shortest path. If the tentative distance from source to the reached node is greater than the upper bound, the reached node is pruned. The process is repeated with v_3, the closest boundary node to v_2, having upper bounds for the shortest-path tree of v_3, and so on. The *centralized shortest-path search* [Hilger et al., 2009, Köhler et al., 2006] improves the preprocessing time costs running a single search from all the boundary nodes b_i to obtain all distances, by storing distances to reached nodes v into an array A, with $A[i] = d(b_i, v)$.

Finally, if the partitions have many boundary nodes, it is possible to take advantage of the PHAST method, speeding up even more this edge-flags preprocessing. The method of PHAST [Delling et al., 2011b] also has a preprocessing phase that is the same used in CH. For this reason, it is only efficient if there are many shortest-path tree computations to amortize this preprocessing cost. First, this method initializes the node distance labels, $d(v) = \infty$, setting $d(s) = 0$. Afterward, the distance label computation is performed in two consecutive processes. The first one runs a Dijkstra search from s to the remaining nodes in $G_\uparrow$ updating the corresponding distance labels of visited nodes. This process stops when there are no more nodes in the priority queue. Then, the second one checks the vertices of $G_\downarrow$ in descending rank order continuing the distance label relaxation.

Applying PHAST to arc-flags precomputation implies running a final process to actually compute which edges of the graphs belong to the shortest-path tree. At this point, all distance labels have been calculated, so the process traverses the edge list of G checking for each edge $e = (u, v)$ whether its weight $w(e) = d(v) - d(u)$. If true, this edge belongs to the shortest-path tree generated from the boundary node of region r. Therefore, the flag of edge e for region r will be set to true.

Experimental results are presented in [Delling et al., 2012], showing how a preprocessing phase of edge flags can be speeded up from several hours, using the naïve Dijkstra algorithm, to a few minutes, using a particular implementation of PHAST, called GPHAST. This implementation takes advantage of the modern graphic processor units (GPUs) computing architectures.

5.2.3   QUERY

The query uses the original graph with an array of flags per edge. The algorithm is a simple modification of Dijkstra's algorithm. At the beginning, the region r_i of the goal node t is retrieved. Every time an edge is considered, the corresponding flag is queried. If the flag is set, then this edge belongs to a shortest path from that node to region r_i, so the algorithm relaxes it. The efficiency of this technique decreases as the search gets closer to the target, because the number of edges with the r_i flag set increases near the boundary area of the target region. This fact changes the nature of the search from depth to breadth, known as the *cone effect*. To mitigate this effect, this technique can be combined with bidirectional search. Note that flags are not designed for backward searches, so the number of flags should be increased to four per edge and region.

When the search is close enough to the target, the query works as a bare Dijkstra search, because all the surrounding edges will have set the flags corresponding to the target region. In

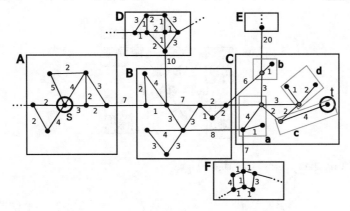

Figure 5.6: Partition into regions of the example graph.

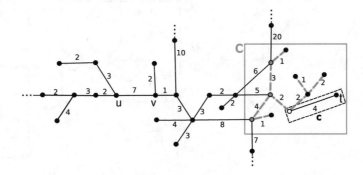

Figure 5.7: Reversed shortest-path trees from boundary nodes of region C. Dashed edges represent the reversed shortest path from the boundary node of second-level region c.

order to improve this step, [Möhring et al., 2005] added a second-level partition to the graph, creating another sub-division for every region. When the search is close enough, it switches to the edge flags of the second-level partition. The creation process of these second-level flags is independent for each region, and each edge will have two flag vectors associated to it.

5.2.4 CASE STUDY

The partition Figure 5.6 shows the partition done for the study case graph. The partition was chosen to simplify the following explanations. We only present region C with a second-level division because it will be the target region of the example. There, every gray node is the boundary node of its second-level region.

Preprocessing In preprocessing phase it is necessary to perform all reversed shortest paths from boundary nodes of every region. In our example, only the reverse shortest paths from the target

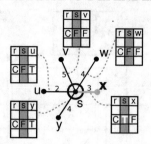

Figure 5.8: Starting query phase of Edge Flags.

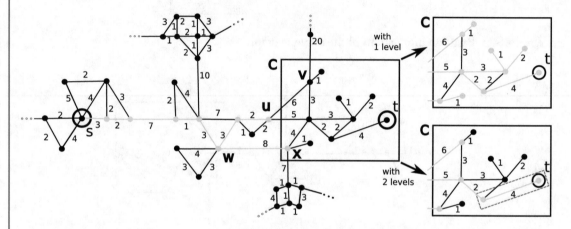

Figure 5.9: Settled nodes from an Edge Flags algorithm query until it reaches the target region (left), inside region C using one level (right, top) and inside region C using two levels (right, bottom).

region and subregion (C and c) are enough to deploy correctly the query. Figure 5.7 shows the reversed shortest-path trees from boundary nodes (dark gray nodes) of region C. The edges of these trees will have the flag to region C set for the head node of each edge. For example, in the edge (u, v), the flag to region C for u ($u\_flag_{(u,v),C}$) will be set. The figure also shows the reversed shortest-path tree (dashed edges) from the boundary node of subregion c. Analogously, these edges will set to true its second-level flags to subregion c for the head node of each edge.

Query   Starting the query phase from source node (see Figure 5.8), it checks the adjacent edges looking for edges whose flag of region C (target region) are set. Edges (s, v) and (s, w) do not have the flag set, so nodes v and w are automatically pruned. Edges (s, u) and (s, y) have the flag set, but it is only for the forward direction from u and y respectively. Therefore, they are also pruned. Finally, edge (s, x) has the flag set and node x is reached. The algorithm continues settling node x and repeating the process.

Figure 5.9 (left) shows all settled nodes of the Edge Flags algorithm query until it reaches the target region. This example shows the typical behavior of the algorithm. At the beginning the search prunes futile nodes and takes the correct path. However, when reaching the target zone, the search is expanded due to the cone effect (edges (u, v) and (w, x) in the figure). The more boundary nodes, the more edges belonging to shortest paths leading to the target region.

Since around the target the algorithm behaves like Dijkstra's algorithm, the search is expanded ever more (see Figure 5.9 (up)). The two-level variant considerably improves this search, going straight to the target node (see Figure 5.9 (down)).

5.3   REACH-BASED ROUTING

Representing the search as a snowball whose size grows with the distance from the source node, we can prune a reached node if we know that the maximum snowball size that this node can handle is smaller than the incoming one. Following this idea, the Reach-Based Routing algorithm [Gutman, 2004] aims to prune those futile reached nodes, depending not only on how far from source location are the nodes but also on how close the target location is. Thus, each node on the graph will have an attribute with an upper bound for both metrics called reach value.

5.3.1   PREPROCESSING: COMPUTING REACH VALUES

The reach value of a node v is computed as follows. A node v in a shortest path $Q = \langle s, .., t \rangle$ has a reach value $r(v, Q)$ that is the minimum value between distance from source, $d(s, v)$, and distance to goal, $d(v, t)$. If node v belongs to more than one shortest path between any pair of nodes, its reach value $r(v, G)$ in a graph G will be the maximum reach value from all the shortest paths that it belongs to (see Fig. 5.10).

Knowing all the reach values of every node implies computing all the possible shortest paths between all pairs of nodes in the graph. In big graphs, such as European or North American road networks, the cost of this step is prohibitive. To find out the reach values without developing an exhaustive preprocessing, a solution is to get a reach upper bound for a node. The idea is to compute first the upper bounds of nodes with a small reach value (for example, nodes near the borders of the graph), and later use them to get the bounds of nodes with a larger reach value, until the reach value of an arbitrary number of nodes have been computed. The remaining nodes will have an infinite reach value. Note that if every node had infinite values, there would be no discards, leading to a bare Dijkstra algorithm.

5.3.2   QUERY

The query phase is a small modification of Dijkstra's algorithm, adding a testing function of the reach value. If $d_e(v, t)$ is as a lower bound of $d(v, t)$ (for example the Euclidean distance, used by [Gutman, 2004]), every time a node v is reached, the condition $d(s, v) > r(v, G)$ and $d_e(v, t) > r(v, G)$ is checked. If true, node v is pruned.

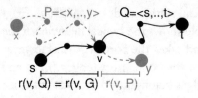

Figure 5.10: Example of different reach values in two shortest paths for a node v.

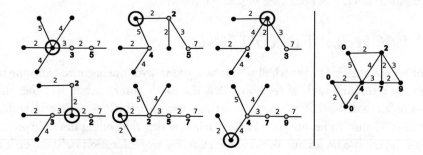

Figure 5.11: Obtaining reach values.

This approach could be combined with other heuristics in order to improve lower bounds of $d_e(v, t)$, or even to apply an edge weight transformation to turn the approach into a goal-directed approach as it is described in Sect. 5.5.

5.3.3   CASE STUDY

We will omit the dark area close to the source node in order to simplify the explanation on how the algorithm works in preprocessing. If the dark area close to source node were not omitted, its preprocessing would change some reach values.

Preprocessing   Computing all possible shortest paths we obtain the exact reach values. Figure 5.11 (right) shows a small portion of the case study near the source node. The left part of the figure represents each shortest path starting at the rounded node. For those shortest paths, bold numbers are their corresponding reach values. The right part of Fig. 5.11 shows the final reach values for the graph, also in bold numbers.

Query   Figure 5.12 (a) shows the first steps of the case study, where the algorithm simply prunes all adjacent nodes to the source node (dashed edges) not involved in the shortest path. Their distance from source node and their estimation to the target one (circle-bounded numbers) are higher than the reach value, thus fulfilling the pruning criteria: $d(s, u) > r(u, G)$ and $d_e(u, t) > r(u, G)$.

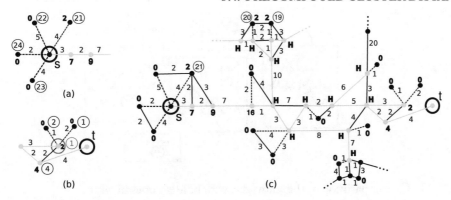

Figure 5.12: (a) First pruning of Reach algorithm. (b) Example of the weakness of the Euclidean distance estimator. (c) Settled nodes from the query phase of the Reach algorithm.

Figure 5.12 (b) shows the behavior of the query near the target. This is a good example of the disadvantages of using the Euclidean estimator, because it forces us to settle a futile node (rounded). The node does not fulfill the pruning criteria because its estimation $d_e(w, t) = 1$ is lower than its reach value, $r(w, G) = 2$.

Figure 5.12 (c) represents all the settled nodes (light gray) in the reach query of the case study. H value means infinite reach value. Nodes with H value belong to a shortest path whose reach value cannot be calculated, since we do not have enough information about the dark areas of the graph. Nodes adjacent to dashed edges are pruned by the reach criteria.

5.4   PRECOMPUTED CLUSTER DISTANCES

The Precomputed Cluster Distances (PCD) approach [Maue et al., 2006, 2010] is based on upper and lower bounds calculated through precomputed distances between clusters of the graph. In the query phase, if a node has a lower bound higher than the current upper bound value of the search, the node is pruned. According to [Delling et al., 2009], its speed-up is comparable to ALT's, but using less space.

5.4.1   PARTITIONING

Among the methods proposed by the authors, the simplest one is *k-center clustering*. k nodes are selected to be used as the center of k different clusters, and the rest of nodes are associated to their closest center, composing a cluster.

5.4.2   PREPROCESSING: COMPUTING UPPER AND LOWER BOUNDS

The objective of this phase is to create a complete distance table with all the minimum distances among the clusters, $d(V_i, V_j) \leftarrow min_{u \in V_i, v \in V_j} d(u, v)$. The nodes u, v that lead to the minimum

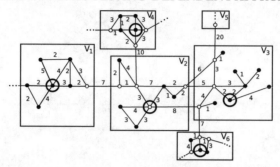

Figure 5.13: Cluster partition of the study case with border nodes in white.

distance are the border nodes, so-called boundary nodes. The table is populated iteratively. For each cluster S, $d(S, V_i)$ is computed for $i = 1, \ldots, k$ using any shortest-path algorithm. Border nodes used are also stored.

5.4.3   QUERY

The algorithm supports unidirectional and bidirectional versions. Only bidirectional search, with a better performance, is explained below. The query is composed by two phases. In the first phase, a bidirectional Dijkstra search is performed from s and t until (a) the searches meet and the algorithm finishes, or (b) until both searches reach the first border nodes s', t' of their regions S, T, and the algorithm continues in the second phase. At this point the search knows $d(s, s')$ and $d(t', t)$.

The *second phase* prunes those nodes w that have a lower-bound estimation of $d(s, t)$ higher than a global upper bound, $\bar{d}(s, w, t) > \hat{d}(s, t)$. This estimation of $d(s, t)$ through w, is computed as $\bar{d}(s, w, t) \leftarrow d(s, w) + d(W, T) + d(t', t)$. Simultaneously, each time the forward search finds a border node u in cluster U stored in precomputation, the upper bound of $d(s, t)$ can be calculated as $\hat{d}(s, t) \leftarrow d(s, u) + d(u, t_{UT}) + d(t_{UT}, t)$.[1] If a new lower $\hat{d}(s, t)$ is computed, the previous one is discarded. The backward search works analogously.

5.4.4   CASE STUDY

As with other algorithms, we will first omit the dark area close to the source node to ease the query phase explanation.

Partition and preprocessing   The partition made for the study case is a synthetic *k-center clustering* (see Figure 5.13). The k nodes have been manually chosen (circle-bounded nodes) and the rest of nodes have been associated to the closest center.

---

[1] If t_{UT} has not been settled in backward search yet, $d(t_{UT}, t)$ is not known, so it is substituted by a pessimistic bound: $2 \times r(T)$, being $r(T)$ the radius distance of cluster T.

Table 5.1: Cluster distances from V_1 and V_3 to the remaining clusters

Clusters	Dist.	Clusters	Dist.
$V_1 \rightarrow V_2$	7	$V_3 \rightarrow V_1$	19
$V_1 \rightarrow V_3$	19	$V_3 \rightarrow V_2$	5
$V_1 \rightarrow V_4$	18	$V_3 \rightarrow V_4$	21
$V_1 \rightarrow V_5$	42	$V_3 \rightarrow V_5$	20
$V_1 \rightarrow V_6$	26	$V_3 \rightarrow V_6$	7

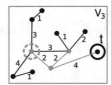

Figure 5.14: First phase of bidirectional PCD algorithm.

Border nodes are represented as white nodes. Distances from one cluster to the rest are computed and stored in tables (see Table 5.1) using any shortest-path algorithm from source-cluster border nodes to remaining border nodes.

Query   Figure 5.14 describes the first phase of the query. The local bidirectional search starts from source s and target t until cluster frontiers are reached (circle-dashed nodes). In forward search, the border node u has been reached. Therefore, the new upper bound for the search will be $\hat{d}(s,t) \leftarrow d(s,u) + d(V_1, V_3) + d(t_{V_1 V_3}, t)$. However, the last term is still unknown, since backward search has not reached node $t_{V_1 V_3}$. Then, this term is estimated using $r(V_3)$, that is 6 for the study case. Therefore, $\hat{d}(s,t) \leftarrow d(s,u) + d(V_1, V_3) + 2 \times r(V_3) = 5 + 19 + 12 = 36$. In the second phase (see Figure 5.15) the pruning criteria $\bar{d}(s,w,t) > \hat{d}(s,t)$ is checked for every reached node w. When the backward search finds node v, the upper bound is updated to 34 because now $d(t_{V_1 V_3}) = 10$ is known. When both searches enter into region V_2 they settle nodes x and y because their lower bound (23) is smaller than the upper one, unlike nodes i, j, and k, that are pruned (resp. 50, 76, and 48). The algorithm finishes when node m is settled by both searches.

If we do not omit the dark area close to the source node, forward search will try to enter into the previously omitted dark area. However, this branch is pruned early when the first upper bound appears. Therefore, the final result of the query phase is not significantly different from the modified one.

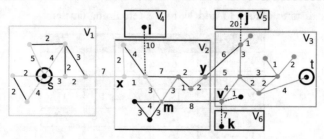

Figure 5.15: Settled nodes of a bidirectional PCD algorithm query.

5.5   OTHER NON-HIERARCHICAL PREPROCESSING METHODS

Separator-based methods [Lipton and Tarjan, 1979] partition the graph into small subgraphs by choosing a set of nodes called *separator nodes*. Preprocessing from each node to a subset of these separator nodes, it is possible to give the exact distance [Fakcharoenphol and Rao, 2006] or an approximation [Muller and Zachariasen, 2007] of any point-to-point query.

A new separator-based method, Customizable Route Planning (CRP) [Delling et al., 2011c, 2013b], allows us to compute shortest paths, by keeping several active metrics at once and generating new ones on the fly. CRP not only supports real-time updates but also increases query speed-ups relative to Dijkstra from fewer than 60 (previous approaches) to 3,000. It divides the preprocessing in two stages: a *metric-independent preprocessing* of the graph topology, that is needed only once; and a *metric customization*, that quickly returns a smaller data space for each metric. Therefore, the query can use the graph real-time information or even also a linear combination of metrics.

The Geometric Containers approach [Schulz et al., 2000, Wagner and Willhalm, 2003, Wagner et al., 2005], that is similar to Edge Flags, precomputes for each edge *e* all nodes that are in a shortest path starting with *e*, storing a geometric shape that contains them. The search prunes the edges whose containers do not contain the target node.

RE [Goldberg et al., 2006], REAL [Goldberg et al., 2006, 2007], and ReachFlags [Bauer et al., 2008], are improved versions of Reach algorithm. The former is a bidirectional variant with more pruning criteria added. The others combine two preprocessing approaches, REAL as a combination of RE and ALT, and ReachFlags as a combination of RE and Edge Flags.

5.6   COMBINING HIERARCHICAL WITH NON-HIERARCHICAL METHODS

Other proposals combine hierarchical with non-hierarchical techniques. The use of separator nodes (see Sect. 5.5) with multilevel techniques allows the creation of a series of interconnected overlay graphs in a hierarchical fashion. CRP [Delling et al., 2011c, 2013b] is an example. HH*

[Delling et al., 2006b] combines the goal-directed technique of ALT with Highway Hierarchies. SHARC [Bauer and Delling, 2008] combines a hierarchical multilevel approach with edge flags (SHortcut + ARC flags). Finally, CALT, CHASE, and TNREF [Bauer et al., 2008] are combinations of Core-Based Routing + ALT, Contraction Hierarchies + Edge Flags, and Transit-Node Routing + Edge Flags, respectively.

CHAPTER 6

Analysis and Comparison of Approaches

In this chapter we will briefly compare the search spaces of the query phase of the approaches described. An in-depth comparison of all algorithms reviewed goes beyond the objectives of this paper, because it must take into account the particular distribution of nodes and edges in the graph, the relative positions of source and target nodes, and the relative cost of each operation involved. An experimental comparison can be found in [Delling et al., 2009], a study that includes needed space, preprocessing times, and query speed-ups. Instead, we will make a quantitative comparison of the search spaces using graphs that follow a triangular tiling pattern with the same weight for the edges. This process allows us to quantify the number of reached and settled nodes of each approach with mathematical formulations, depending on the distance between the source and destination nodes. The reader should take into account, however, that our choice of graphs for this comparison constitutes a simplification, since most of the networks of the real world do not follow strict patterns, being more irregular. Later, we will show a qualitative analysis and comparison for more irregular graphs than the triangular tiling just assuming a homogeneous distribution of nodes and edges.

6.1 QUANTITATIVE SEARCH SPACE ANALYSIS

In this section are shown the worst and best cases that a query can find in terms of reached and settled nodes.

6.1.1 DIJKSTRA'S ALGORITHM

The algorithm starts relaxing the edges of the source node reaching all adjacent nodes. Note that these nodes conform a hexagon around the source because they are at the same distance to it. At the following iteration, it arbitrarily selects one of these reached nodes to settled it. Later, it performs the edge-relaxing operation on the new settled node reaching new nodes with higher distance value than the previous reached ones. The algorithm then will proceed settling all concentric reached nodes that have the same tentative distance, and therefore reaching nodes that conform a bigger hexagon. We define as $\mathbb{H}(d)$ the number of nodes that are inside a regular hexagon centered at source node with radius d, including the nodes belonging to its perime-

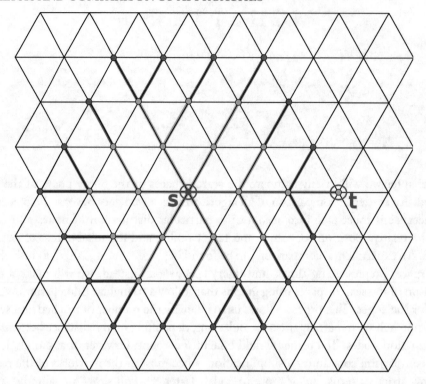

Figure 6.1: Best case of a Dijkstra search in a trihexagonal grid. Reached nodes are depicted in red and settled nodes in green, whereas source node (s) and target node (t) are colored in blue.

ter. Note that this value excludes the source vertex unlike the *hex number* widely known by the community.

$$\mathbb{H}(d) = 2 \cdot \left(d + \sum_{i=1}^{d} (d+i) \right) = 2 \cdot \left(d^2 + d + \sum_{i=1}^{d} i \right) = 6 \cdot \sum_{i=1}^{d} i = 3d^2 + 3d \qquad (6.1)$$

Best case: The situation where the algorithm carries out the lowest number of iterations occurs when it selects the target node as the first node of the list of reached vertices with the same tentative distance. Note that in this situation, the algorithm has to settle all concentric hexagonal rings of nodes with lower tentative distance than the target. Figure 6.1 shows the pattern of the search space for the best case. Following this behavior, the number of both reached ($\mathbb{R}_b$) and settled nodes ($\mathbb{S}_b$) can be mathematically calculated depending the shortest-path distance (d) through the following equations:

$$\mathbb{R}_b(d) = \mathbb{H}(d) = 3d^2 + 3d \qquad (6.2)$$

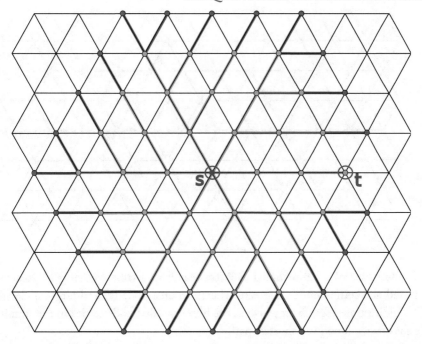

Figure 6.2: Worst case of a Dijkstra search in a trihexagonal grid. Reached nodes are depicted in red and settled nodes in green, whereas source node (s) and target node (t) are colored in blue.

$$\mathbb{S}_b(d) = \mathbb{H}(d-1) + 1 = 3(d-1)^2 + 3(d-1) + 1 = 3d^2 + 3d - 5 \qquad (6.3)$$

Worst case: The situation where the algorithm carries out the highest number of iterations occurs when it selects the target node as the last node of the list of reached vertices with the same tentative distance. Note that in this situation, the algorithm has to settle all nodes inside the concentric hexagonal rings with radius equal to the target tentative distance. Figure 6.2 shows the pattern of the search space for the worst case, and the equations for the reached ($\mathbb{R}_w$) and settled nodes ($\mathbb{S}_w$) are the following:

$$\mathbb{R}_w(d) = \mathbb{H}(d+1) - 1 = 3(d+1)^2 + 3(d+1) + 1 = 3d^2 + 9d + 5 \qquad (6.4)$$

$$\mathbb{S}_w(d) = \mathbb{H}(d) = 3d^2 + 3d \qquad (6.5)$$

6.1.2   BIDIRECTIONAL DIJKSTRA'S ALGORITHM

The algorithm starts relaxing the edges of the source node reaching all adjacent nodes conforming the hexagon around the source. At the following iteration, the bidirectional version selects the target to continue the search in a backward way. Then the process is repeated using the incoming

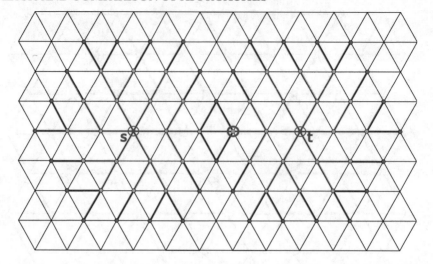

Figure 6.3: Best case of a bidirectional Dijkstra search in a trihexagonal grid. Reached nodes are depicted in red and settled nodes in green, whereas source node (s) and target node (t) are colored in blue. The bidirectional search stops when the blue-rounded node settles in the middle of the grid because it has already found the shortest path.

edges instead of the outgoing ones, and it reaches the nodes conforming the hexagon around the target. Later it continues performing the edge-relaxing operation alternating between forward and backward searches basing on a specific policy. The algorithm then will proceed settling all concentric reached nodes that have the same tentative distance around the source and the target, and therefore reaching nodes that conform two hexagons.

Best case: The situation where the algorithm carries out the lowest number of iterations occurs when it selects a previously settled node by the opposite search as the first node of the list of reached vertices with the same tentative distance. The algorithm has to settle all concentric nodes of the two hexagonal rings with lower tentative distance than the half shortest-path distance between source and target. Figure 6.3 shows the pattern of the search space for the best case, and the following equations calculate the number of reached ($\mathbb{R}_b$) and settled nodes ($\mathbb{S}_b$):

$$\mathbb{R}_b(d) \;=\; 2 \cdot \mathbb{H}\left(\lceil d/2 \rceil\right) - (d\%2)(\lceil d/2 \rceil - 1)^2 \tag{6.6}$$

$$=\; 6\lceil d/2 \rceil^2 + 6\lceil d/2 \rceil - (d\%2)(\lceil d/2 \rceil^2 - 1) \tag{6.7}$$

$$\mathbb{S}_b(d) \;=\; 2 \cdot \mathbb{H}\left(\lfloor (d-1)/2 \rfloor\right) + ((d-1)\%2) \tag{6.8}$$

$$=\; 6\lfloor (d-1)/2 \rfloor^2 + 6\lfloor (d-1)/2 \rfloor + ((d-1)\%2) \tag{6.9}$$

Worst case: The situation where the algorithm carries out the highest number of iterations occurs when it selects the previously settled node by the opposite search as the last node of the list

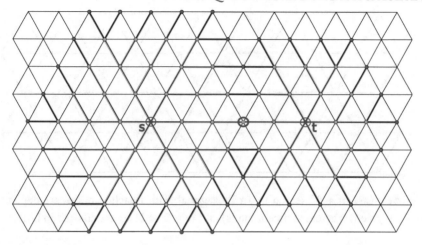

Figure 6.4: Worst case of a bidirectional Dijkstra search in a trihexagonal grid. Reached nodes are depicted in red and settled nodes in green, whereas source node (s) and target node (t) are colored in blue. The bidirectional search stops when the blue-rounded node settles in the middle of the grid because it has already found the shortest path.

of reached vertices with the same tentative distance. Note that in this situation, the algorithm has to settle all nodes inside the concentric hexagonal rings with radius equal to the target tentative distance. Figure 6.4 shows the pattern of the search space for the worst case, and the equations for the reached ($\mathbb{R}_w$) and settled nodes ($\mathbb{S}_w$) are the following:

$$
\begin{aligned}
\mathbb{R}_w(d) &= \mathbb{H}\left(\lceil d/2 \rceil + 1\right) + \mathbb{H}\left(\lfloor d/2 \rfloor + 1\right) - \left(\lceil d/2 \rceil\right)^2 & (6.10) \\
&= 2\lceil d/2 \rceil^2 + 3\lfloor d/2 \rfloor^2 + 9\left(\lceil d/2 \rceil + \lfloor d/2 \rfloor\right) + 12 & (6.11)
\end{aligned}
$$

$$
\begin{aligned}
\mathbb{S}_w(d) &= \mathbb{H}\left(\lceil d/2 \rceil\right) + \mathbb{H}\left(\lfloor d/2 \rfloor\right) - 1 & (6.12) \\
&= 3\left(\lceil d/2 \rceil\right)^2 + 3\left(\lceil d/2 \rceil\right) + 3\left(\lfloor d/2 \rfloor\right)^2 + 3\left(\lfloor d/2 \rfloor\right) - 1 & (6.13)
\end{aligned}
$$

6.1.3 A* ALGORITHM

The algorithm starts relaxing the edges of the source node reaching all adjacent nodes. In each relaxing process it has to compute the new weight for the edge. In this case the nodes of the hexagon around the source do not have the same tentative distance. The closer the node is to the target, the lower is its tentative distance in the priority queue. Note that if the reached node is in the middle of the straight line to the target it will have the lowest tentative distance and it will be the first to be chosen in the next step. Then the algorithm chooses one of these closest nodes to settle it at the following iteration, and performs the edge-relaxing operation modifying the

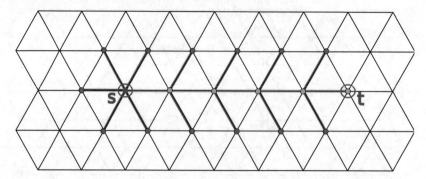

Figure 6.5: Best case of a A* search in a trihexagonal grid. Reached nodes are depicted in red and settled nodes in green, whereas source node (s) and target node (t) are colored in blue.

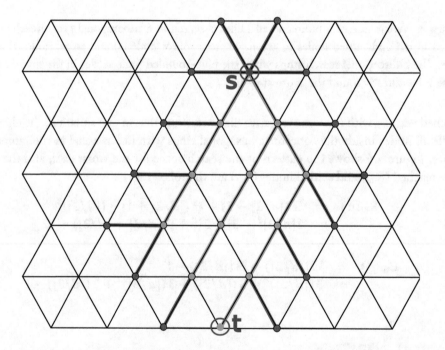

Figure 6.6: Worst case of a A* search in a trihexagonal grid. Reached nodes are depicted in red and settled nodes in green, whereas source node (s) and target node (t) are colored in blue.

weights of the new edges. The algorithm then will proceed settling all reached nodes in the line or in the closest hexagon to the target.

Best case: The algorithm carries out the lowest number of iterations when the source and target node are connected through a straight line. In this situation, the algorithm only settles those nodes that belong to the line, reaching only their adjacent vertices. Figure 6.5 shows the pattern of the search space for this case. The number of both reached ($\mathbb{R}_b$) and settled nodes ($\mathbb{S}_b$) is calculated through the following formulations:

$$\mathbb{R}_b(d) = 3 \cdot (d + 1) \tag{6.14}$$

$$\mathbb{S}_b(d) = d \tag{6.15}$$

Worst case: In the trihexagonal grid, the worst case for the A* algorithm occurs when the source and the target are vertically separated, as it is shown in Fig. 6.6. There the algorithm has several nodes that have the same tentative distance to the target unlike the best case that always has only one node (the one in the straight line). Both reached ($\mathbb{R}_w$) and settled nodes ($\mathbb{S}_w$) form a diamond shape, and these corresponding values are computed by:

$$\begin{aligned}
\mathbb{R}_w(d) &= (\lfloor d/2 \rfloor)^2 + (d\%2)(7\lfloor d/2 \rfloor + 7) + ((d+1)\%2)(3d + 4) & (6.16)\\
&= (\lfloor d/2 \rfloor)^2 + 6\lfloor d/2 \rfloor + 4 + (d\%2)(\lfloor d/2 \rfloor + 3) & (6.17)
\end{aligned}$$

$$\begin{aligned}
\mathbb{S}_w(d) &= (\lceil d/2 \rceil)^2 + (d\%2)(\lceil d/2 \rceil - 1) + ((d+1)\%2)d & (6.18)\\
&= (\lceil d/2 \rceil)^2 + \lceil d/2 \rceil - (d\%2) + ((d+1)\%2)(d/2) & (6.19)
\end{aligned}$$

6.1.4   TRENDS AND COMPARISONS

Figures 6.7 and 6.8 show the trends for the studied approaches in terms of reached and settled nodes respectively. We can observe that for computing the shortest-path distance between two nodes that are very close, $d \leq 5$, both methods, Dijkstra's algorithm and its bidirectional variant, explore almost the same search space. As the distance between source and target increases, the search space of the bidirectional variant becomes significantly smaller than Dijkstra's algorithm. Note that carrying out the bidirectional algorithm is needed to have an inverse representation of the graph for the backward search, and the mechanisms to handle the two searches simultaneously. Therefore, this reduction in the search space implies a reduction in terms of temporal costs because fewer iterations are needed to complete the query at the cost of more complex programming and higher space costs. The same happens with the A* approach. It explores a more reduced search space at the cost of computing in each edge-relaxing operation the new values for the weights. As the distance between the two involved nodes increases, the effort of computing the heuristic becomes more profitable because it avoids exploring huge search spaces. Note that in the case of computing shortest-path distances in road maps, a good heuristic is to calculate the Euclidean distance, that is not a very costly computation in terms of execution time, but another heuristic with a higher computational cost could ruin the benefits of applying the approach.

The preprocessing-dependent approaches are not contemplated in the trihexagonal region because most of them do not achieve improvements over the classical approaches since the topology of the graph is regular. The cardinality computation of the reached/settled sets for these approaches implies taking into account more parameters that are particular to each method (as

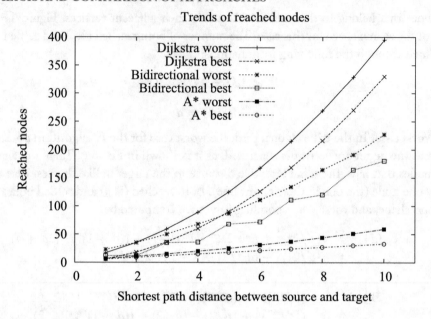

Figure 6.7: Trends for the reached nodes in the best and worst cases.

neighbor cardinality for HH or partition size for EF among others), and therefore its addition would lead to an unfair quantitative comparison. The strengths of the preprocessing-dependent approaches consist of obtaining some significant information of an irregular graph before performing the search, in order to reduce the query time whenever possible.

6.2   QUALITATIVE SEARCH SPACE ANALYSIS

In graphs that do not follow a regular topology or pattern it is more difficult to predict what the algorithm can find after selecting one node to be settled, or even if it will waste the effort of exploring a path due to a dead end. The preprocessing phase obtains information that will be useful in the query phase in order to direct the search toward the destination or with the aim to prune useless paths. In the following sections we will compare the weaknesses and strengths of each of these methods studied through the synthetic case study we have designed for, and what is the search space behavior of the approaches through a graphical depiction in graphs with a non-regular homogeneous node distribution.

6.2.1   LESSONS FROM OUR CASE STUDY

The application of the classical methods to the synthetic case study leads, through the use of an irregular example, to similar conclusions as the analysis made in the quantitative section through the mathematical formulations. The search space of Dijkstra's algorithm grows settling all nodes

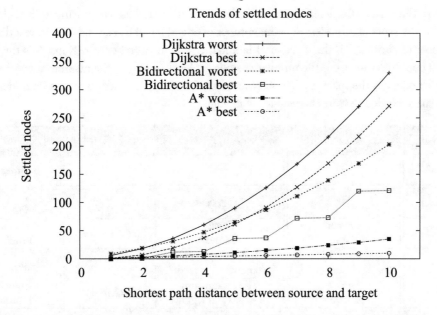

Figure 6.8: Trends for the settled nodes in the best and worst cases.

inside the ball radius, turning out a costly method in the most dense areas because no pruning criteria is deployed. Figure 3.3 shows this behavior where the algorithm settles all nodes of the graph and more not depicted in the dark areas before reaching the target node. The application of a bidirectional variant of this method reduces the size of this search space to less than the half at the cost of duplicating the space cost (an additional reverse graph is needed), and a more complex policy for the priority queue (a forward and a backward search are performed). In Fig. 3.4 we see that the settling radius is reduced to the half, leaving out the top and bottom node areas, but settling all nodes inside of the forward and backward search balls due to the lack of pruning criteria.

The application of heuristics, as the Euclidean distance, in the A* algorithm allows the search to select a correct node to be settled, giving more priority to those nodes that are closer to the target with the hope of finding the destination before exploring the zones the bidirectional Dijkstra settles. This method in the study case (see Fig. 3.7) reduces the exploration in the dark area located at the left of the source node, and also avoids the search to enter in the top and bottom node areas. However, heuristics usually do not consider all cases, and in some cases it could ruin the algorithm performance. For example, an algorithm that uses the Euclidean distance for road maps cannot take into account if there is a landform in the middle of the closest node to the target and the target itself, or even if the topology of the possible road that connects them is not straight (curves, inclination, ...). In the study case we can observe this weakness when the algorithm selects the two nodes with $h(v) = 1$ instead of taking the correct one at the bottom

Table 6.1: Summary of different features of the methods described, comparing with Dijkstra's algorithm. n is the number of nodes; m is the number of edges; δ is the space used to store a distance; k is the number of clusters; l is the number of landmarks; r is the number of regions; β is the space used to store a boolean value; n_L is the number of nodes at level L; t_{max} is the maximum number of access nodes per node; ε is the space used to store an edge; λ is the space used to store a level identifier; and υ is the space needed to store the weight of an edge

	Stage needed before preprocessing	Preprocessing result	Preproc. spatial complexity	Mechanism that speeds up query
A*	-	-	-	Reordering
Reach	-	Reach value	$\Theta(n\delta)$	Geometric and bound Pruning
PCD	Cluster partition	Cluster distances	$\Theta(k^2\delta - k\delta)$	Bound Pruning
ALT	Landmark selection	Distances from nodes to landmarks and back	$\Theta(2nl\delta)$	Reordering
EF	Region partition	SP relationships of edges and regions	$\Theta(2mr\beta)$	Geometric Pruning
HH	-	Shortcuts	$O(m(\varepsilon + \lambda))$	Hierarchical Pruning
HH-DT	-	Shortcuts and distances between all n_L nodes	$O(m(\varepsilon + \lambda)+$ $+(n_L^2\delta - n_L\delta))$	H. Pruning and Table lookups
CH	Order selection	Node levels and shortcuts	$O(n^2(\varepsilon + \delta)+$ $+n(\lambda - \varepsilon - \delta))$	Order-base H. Pruning
TNR	Transit-Node set computing	Distances	$O(2nt_{max}\delta+$ $+T^2\delta - T\delta)$	Table lookups
HL	-	Labels	$O(n^2(\upsilon + \delta)+$ $+n(\upsilon - \delta))$	Label computation

with $h(v) = 4$. The combination of both approaches, A* and bidirectional search, does not turn out in an improved approach for the study case (see Fig. 3.8), and its implementation results are very complex due to the combination of a forward and a backward heuristics in addition to the intrinsic difficulties of the bidirectional search. Although the A* algorithm seems to be a good choice, it can be noticeably improved if we can deploy a preprocessing phase that allows the algorithm to better direct the search toward the destination, to deploy a pruning criteria, or to create a hierarchy of the graph.

The preprocessing phase of Landmark A* (ALT) algorithm computes shortest-path distances from all nodes to a set of vertices called landmarks and vice versa. With this information, the algorithm is able to calculate better lower bounds for the heuristic values of the A* algorithm, trying to better direct the search toward the target. In this way, this improved version does not have the weaknesses derived from the use of the Euclidean distance because it is based on the

Table 6.2: Weaknesses and strengths of the different approaches. Some methods rely on the "quality" of the results of a phase previous to the precomputation. Some examples are a good ordering in CH, or a good partitioning in EF. The Dark Area Avoiding criterion for hierarchical methods is not always fulfilled, but, in that cases, the cost of traversing them is small compared with the other methods. Note that all non-hierarchical methods can be also modified to perform a Bidirectional Dijkstra, but they were not originally designed for this variant, as the hierarchical ones. The use of HH or CH in TNR is only to solve local queries

	Dark area avoiding	Obstacle avoiding	Recover paths	Good for	Relies on a pre-preproc.	SP alg. used in preproc.	SP method for query
Dijkstra	×	×	×	Near	×	×	-
A*	×	×	×	Near	×	×	Dijkstra
Reach	×	×	×	Near	×	×	Dijkstra
PCD	×	✓	×	Near	✓	×	Dijkstra
ALT	✓	×	×	Near	✓	×	A*
EF	✓	✓	×	Far	✓	×	Dijkstra
HH	~	✓	✓	Far	×	×	BiDijkstra
CH	~	✓	✓	Far	✓	×	BiDijkstra
TNR	~	✓	✓	Far	✓	HH|CH|Sep.	HH|CH
HL	~	✓	✓	Far	✓	CH	-

weights of the graph instead of the straight line from the node and the target. However, this method needs to have a set of landmarks big enough to achieve a good performance. Figure 5.4 shows the query of ALT algorithm executed in our study case with only one landmark. There the search explores the complete graph even without avoiding the dark areas, that is worse than the search space explored by the crude A* algorithm with the Euclidean distance. On the other hand, if we use a more complete set of landmarks, correctly located in the graph, the search space of ALT algorithm is reduced compared with the A* algorithm. The lower bounds of the heuristic are more reliable, and the correct situation of the landmarks allows the search to avoid entering into any dark area (see Fig. 5.5).

Another approach that tries to better direct the search is Edge Flags. The preprocessing phase labels which edges belong to shortest paths toward a region of the graph. Then the algorithm in the query phase does not need to compute any heuristics to know which path is better, but just checking a flag that indicates if the corresponding edge belongs to a shortest path to the region where the target belongs. This approach is very strong when the queries compute paths to far locations because there is only one path to reach the target region. However, as the search gets closer to this region, the number of shortest path increased, creating a conic search space. Another weakness of this approach is that, when the algorithm reaches the target region, there is no more information about shortest paths, and an exhaustive exploration must be done (using, for example, a Dijkstra algorithm). Both weaknesses can be alleviated by creating smaller regions of the graph, leading to a smaller conic effect, and a reduced search inside the region, but this solution implies having more flag information to store. The application of the second-level partition solves the second weakness with a low space cost and leads to a smaller search space inside the region. Figure 5.9 shows both variants. For the study case, it is possible to see that, although the Edge Flag approach has to deal with the conic effect, the search is better directed than those performed by A*, ALT, Reach, or PCD algorithms, considerably reducing the number of settled nodes.

The first hierarchical approach, Highway Hierarchies, first deploys a naïve Dijkstra inside the so-called neighbors of both source and target nodes. The search continues in the following levels of the subgraph hierarchy, leaving out those nodes that belong to the lower levels (see Fig. 4.4). Although effective, the preprocessing and query phase of this method are complex to implement. Later this method was outperformed and simplified by the Contraction Hierarchies approach, that gives to each node of the graph its own level inside the hierarchy. Then the algorithm only has to continue the search through those nodes with a higher level, instead of exploring all vertices of a higher-level subgraph. Figure 4.9 reflects this significant reduction of the search space to a dozen nodes for the study case.

Finally, the Hub-based labeling algorithm turns out to be the fastest method known. It avoids performing any search by just intersecting the path information of both source and target node, obtained in the preprocessing phase. This phase of precomputation is extremely costly in terms of time, but the use of Contraction Hierarchies to obtain the path data has made it feasible. Even so, the spatial cost for this approach is also very high due to the great amount of information that has to be stored for each node.

6.2.2   UNIFORM SEARCH SPACE COMPARISON

Tables 6.1 and 6.2 summarize some of the features, weaknesses, and strengths of the methods described. To complement this information, the interested reader can find in [Sommer, 2014] a figure comparing the query times with respect to the space requirements for road-networks route-planning algorithms (see Fig. 4 of that work).

As a final summary, Fig. 6.9 intends to give a graphical idea of the search spaces of the methods studied. The general behavior of each method described is the following:

Dijkstra: The exploration space of Dijkstra's algorithm would be a circle with radius $d(s,t)$.

BiDi: In Bidirectional Dijkstra's, the exploration space is reduced to two circles around source and target, with radiuses of approximately half the length of $d(s,t)$.

A*: The algorithm that uses the A* heuristic postpones nodes that are far from the target with the hope of finding the shortest path before settling them. This leads to an ellipse-shaped exploration space.

BiA*: The combination of both bidirectional and A* search considerably reduces the search space, tending to focus the search into two ellipses.

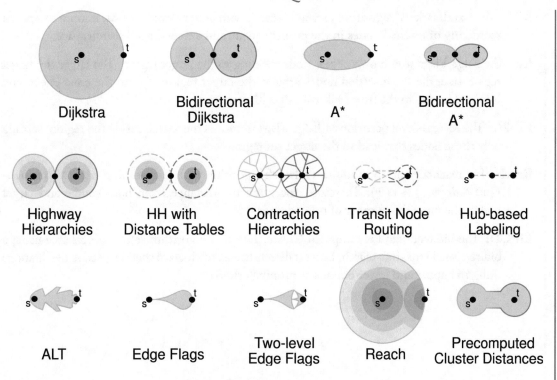

Figure 6.9: Idealized comparison of settled nodes search space for the different algorithms.

HH: Highway Hierarchies reduces the number of settled nodes because it has a bidirectional nature. Moreover, the search is limited by the neighborhood-size parameter in each level. The degraded colors in the figure reflect the relative number of nodes explored in each level.

HH-DT: The use of Distance Tables in Highway Hierarchies speeds up the search in the last level, since only a few table look-ups have to be done.

CH: Contraction Hierarchies runs a bidirectional Dijkstra search, traversing the graph through those nodes with higher rank than their predecessor.

TNR: Transit Node Routing just performs the Distance Table method from the beginning of the query, looking up in tables for the connection between source and target nodes through their access points.

HL: Hub-based labeling just intersects the labels from source and target node, obtaining the shortest-path distance and the intermediate node to recover the path.

ALT: The Landmark-A* algorithm uses a A* search with better lower bounds, leading to a better reordering of reached nodes in the priority queue and a tree-shaped search space.

EF: The Edge Flags search only settles nodes leading to the target region. The larger the target region size, the more settled nodes around the target region, due to the cone effect, and inside of it, due to the bare Dijkstra's algorithm used.

EF-2L: The second-level partition of Edge Flags improves the search inside the region, settling only those nodes that lead to the target subregion.

Reach: The reach algorithm modifies Dijkstra's algorithm with the pruning criteria $d(s, w) > r(w) \wedge d_e(w, t) > r(w)$. Therefore, the more the distance from both source and target nodes, the more the number of pruned nodes.

BiPCD: The bidirectional Precomputed Cluster Distance algorithm settles a local area using a bidirectional Dijkstra's search. Later it directs the search toward their respective destinations using the upper and lower bounds as pruning criteria.

CHAPTER 7

Conclusions

The aim of the quantitative and qualitative analysis carried out is to give some hints to choose an appropriate method depending on each particular situation. For example, if the graph of the network is small, the logical choice is to implement a naïve Dijkstra's algorithm, due to its simplicity and easy understanding. As the size of the network increases, the search space of the queries becomes bigger, and better algorithms may be needed. If the network follows a regular pattern and it is known in advance, in most cases it is possible to find a heuristic for the A^* algorithm that significantly reduces the search space and query time without getting involved in complex programming for preprocessing-dependent approaches. For huge graphs, as the European roadmap with more than 18 million nodes, preprocessing approaches are needed for fast queries, in the order of microseconds. Hierarchical approaches, such as Contraction Hierarchies or Transit-Node Routing, have turned out to solve the queries in this order of magnitude, leaving out for the literature more complex and outperformed approaches such as Highway Hierarchies, Landmark A^*, Reach-based, or Precomputed Cluster Distances. Finally, the combination of Contraction Hierarchies or Transit-Node Routing with the Edge Flags method has created, together with the Hub-based labeling method, the fastest approaches known so far to solve the query phase, at the cost of a complex implementation, high preprocessing times, and elevated data storage costs.

Bibliography

Party: A partitioning library, 2006. `http://www2.cs.uni-paderborn.de/cs/robsy/party.html`. 35

I. Abraham, D. Delling, A. V. Goldberg, and R. F. Werneck. A hub-based labeling algorithm for shortest paths in road networks. In *SEA'11*, pages 230–241, Berlin, 2011. Springer. ISBN 978-3-642-20661-0. DOI: 10.1007/978-3-642-20662-7_20. 27, 29

I. Abraham, D. Delling, A. V. Goldberg, and R. F. Werneck. Hierarchical hub labelings for shortest paths. In *ESA'12*, pages 24–35, Berlin, 2012. Springer. DOI: 10.1007/978-3-642-33090-2_4. 2, 29

J. Arz, D. Luxen, and P. Sanders. Transit Node Routing Reconsidered. In V. Bonifaci, C. Demetrescu, and A. Marchetti-Spaccamela, editors, *Experimental Algorithms*, volume 7933 of *Lecture Notes in Computer Science*, page 55–66. Springer Berlin Heidelberg, 2013. ISBN 978-3-642-38526-1. DOI: 10.1007/978-3-642-38527-8. 30

J. Barceló, E. Codina, J. Casas, J. L. Ferrer, and D. García. Microscopic traffic simulation: A tool for the design, analysis and evaluation of intelligent transport systems. *J. Intell. Robot. Syst.*, 41:173–203, 2005. ISSN 0921-0296. DOI: 10.1007/s10846-005-3808-2. 1

C. Barrett, R. Jacob, and M. Marathe. Formal-language-constrained path problems. *SIAM J. Comput.*, 30:809–837, 2000. DOI: 10.1137/S0097539798337716. 1

H. Bast, S. Funke, and D. Matijevic. Transit—ultrafast shortest-path queries with linear-time preprocessing. In Demetrescu et al. [2006]. 25

H. Bast, S. Funke, D. Matijevic, P. Sanders, and D. Schultes. In transit to constant time shortest-path queries in road networks. In *ALENEX'07*, pages 46–59, Philadelphia, 2007. SIAM. 25, 26, 27, 29

H. Bast, E. Carlsson, A. Eigenwillig, R. Geisberger, C. Harrelson, V. Raychev, and F. Viger. Fast routing in very large public transportation networks using transfer patterns. In *ESA'10*, pages 290–301, Berlin, 2010. Springer. ISBN 3-642-15774-2. DOI: 10.1007/978-3-642-15775-2_25. 1

H. Bast, D. Delling, A. Goldberg, M. Müller-Hannemann, T. Pajor, P. Sanders, D. Wagner, and R. Werneck. Route Planning in Transportation Networks. Technical Report MSR-TR-

64 BIBLIOGRAPHY

2014-4, January 2014. `http://research.microsoft.com/apps/pubs/default.aspx?id=207102`. 2

R. Bauer and D. Delling. Sharc: Fast and robust unidirectional routing. In *ALENEX'08*, pages 13–26, Philadelphia, 2008. SIAM. DOI: 10.1145/1498698.1537599. 45

R. Bauer, D. Delling, and D. Wagner. Experimental study on speed-up techniques for timetable information systems. In C. Liebchen, R. K. Ahuja, and J. A. Mesa, editors, *ATMOS'07*, pages 209–225, Dagstuhl, 2007. IBFI. ISBN 978-3-939897-04-0. `http://drops.dagstuhl.de/opus/volltexte/2007/1169`. 1

R. Bauer, D. Delling, P. Sanders, D. Schieferdecker, D. Schultes, and D. Wagner. Combining hierarchical and goal-directed speed-up techniques for Dijkstra's algorithm. In *WEA'08*, pages 303–318, Berlin, 2008. Springer. DOI: 10.1007/978-3-540-68552-4_23. 44, 45

C. Böhm, E. Kny, B. Emde, Z. Abedjan, and F. Naumann. Sprint: ranking search results by paths. In *EDBT/ICDT'11*, pages 546–549, New York, 2011. ACM. ISBN 978-1-4503-0528-0. DOI: 10.1145/1951365.1951437. 1

J. A. Bondy and U. S. R. Murty. *Graph theory with applications*, volume 6. Macmillan, London, 1976. 3

C. Chen and M. Lee. Global path planning in mobile robot using omnidirectional camera. In *CECNet'11*, pages 4986–4989, Washington, april 2011. IEEE Computer Society. DOI: 10.1109/CECNET.2011.5768666. 1

T. H. Cormen, C. Stein, R. L. Rivest, and C. E. Leiserson. *Introduction to Algorithms*. McGraw-Hill Higher Education, Burr Ridge, Il 60521, 2nd edition, 2001. ISBN 0070131511. 6

G. Dantzig. *Linear Programming And Extensions*. Princeton University Press, Princeton, 1963. 9

D. Delling, M. Holzer, K. Müller, F. Schulz, and D. Wagner. High-performance multi-level graphs. In Demetrescu et al. [2006]. 25

D. Delling, P. Sanders, D. Schultes, and D. Wagner. Highway hierarchies star. In Demetrescu et al. [2006]. 45

D. Delling, P. Sanders, D. Schultes, and D. Wagner. Engineering route planning algorithms. In *Algorithmics of Large and Complex Networks*, pages 117–139, Berlin, 2009. Springer. ISBN 978-3-642-02093-3. DOI: 10.1007/978-3-642-02094-0_7. 7, 41, 47

D. Delling, A. Goldberg, I. Razenshteyn, and R. Werneck. Graph partitioning with natural cuts. In *Parallel Distributed Processing Symposium (IPDPS), 2011 IEEE International*, pages 1135–1146, may 2011a. DOI: 10.1109/IPDPS.2011.108. 35

D. Delling, A. V. Goldberg, A. Nowatzyk, and R. F. Werneck. Phast: Hardware-accelerated shortest path trees. In *IPDPS'11*, pages 921–931, Washington, 2011b. IEEE Computer Society. ISBN 978-0-7695-4385-7. DOI: 10.1109/IPDPS.2011.89. 36

D. Delling, A. V. Goldberg, T. Pajor, and R. F. Werneck. Customizable route planning. In *SEA'11*, pages 376–387, Berlin, 2011c. Springer. ISBN 978-3-642-20661-0. DOI: 10.1007/978-3-642-20662-7_32. 44

D. Delling, A. V. Goldberg, A. Nowatzyk, and R. F. Werneck. Phast: Hardware-accelerated shortest path trees. *Journal of Parallel and Distributed Computing*, 2012. ISSN 0743-7315. DOI: 10.1016/j.jpdc.2012.02.007. 36

D. Delling, A. Goldberg, and R. Werneck. Hub Label Compression. In V. Bonifaci, C. Demetrescu, and A. Marchetti-Spaccamela, editors, *Experimental Algorithms*, volume 7933 of *Lecture Notes in Computer Science*, page 18–29. Springer Berlin Heidelberg, 2013a. ISBN 978-3-642-38526-1. DOI: 10.1007/978-3-642-38527-8. 2

D. Delling, A. V. Goldberg, T. Pajor, and R. F. Werneck. Customizable route planning in road networks. In *Sixth Annual Symposium on Combinatorial Search*, 2013b. 44

C. Demetrescu, A. V. Goldberg, and D. S. Johnson, editors. *9th DIMACS Implementation Challenge - Shortest Paths*, Providence, 2006. American Mathematical Society. 63, 64, 67, 68

E. W. Dijkstra. A note on two problems in connexion with graphs. *Numerische Mathematik*, 1: 269–271, 1959. ISSN 0029-599X. DOI: 10.1007/BF01386390. 5

D. Dreyfus. An appraisal of some shortest path algorithms. Technical Report RM-5433, Rand Corporation, Santa Monica, 1967. 9

J. Fakcharoenphol and S. Rao. Planar graphs, negative weight edges, shortest paths, and near linear time. *J. Comput. Syst. Sci.*, 72:868–889, August 2006. ISSN 0022-0000. DOI: 10.1016/j.jcss.2005.05.007. 44

M. L. Fredman and R. E. Tarjan. Fibonacci heaps and their uses in improved network optimization algorithms. *J. ACM*, 34:596–615, July 1987. ISSN 0004-5411. DOI: 10.1145/28869.28874. 6

C. Gavoille, D. Peleg, S. Pérennes, and R. Raz. Distance labeling in graphs. In *SODA'01*, pages 210–219, Philadelphia, 2001. SIAM. ISBN 0-89871-490-7. DOI: 10.1145/365411.365447. 28

R. Geisberger. Contraction hierarchies: Faster and simpler hierarchical routing in road networks. Master's thesis, University of Karlsruhe, 2008. DOI: 10.1007/978-3-540-68552-4_24. 20, 21, 25

R. Geisberger, P. Sanders, D. Schultes, and D. Delling. Contraction hierarchies: faster and simpler hierarchical routing in road networks. In *WEA'08*, pages 319–333, Berlin, 2008. Springer. ISBN 3-540-68548-0, 978-3-540-68548-7. DOI: 10.1007/978-3-540-68552-4_24. 19, 22, 29

R. Geisberger, P. Sanders, D. Schultes, and C. Vetter. Exact routing in large road networks using contraction hierarchies. *Transportation Science, Articles in Advance*, pages 1–17, August 2011. DOI: 10.1287/trsc.1110.0401. 25

A. V. Goldberg and C. Harrelson. Computing the shortest path: A search meets graph theory. Technical Report MSR-TR-2004-24, Microsoft Research, Vancouver, Canada, 2004. http://research.microsoft.com/apps/pubs/default.aspx?id=64511. 6, 9, 10, 32

A. V. Goldberg and C. Harrelson. Computing the shortest path: A search meets graph theory. In *SODA'05*, pages 156–165, Philadelphia, 2005. SIAM. ISBN 0-89871-585-7. http://portal.acm.org/citation.cfm?id=1070432.1070455. 11, 31

A. V. Goldberg, H. Kaplan, and R. F. Werneck. Reach for A*: Efficient point-to-point shortest path algorithms. In *ALENEX'06*, pages 129–143, Philadelphia, 2006. SIAM. 44

A. V. Goldberg, H. Kaplan, and R. F. Werneck. Better landmarks within reach. In *WEA'07*, pages 38–51, Berlin, 2007. Springer. ISBN 978-3-540-72844-3. DOI: 10.1007/978-3-540-72845-0_4. 32, 44

R. J. Gutman. Reach-based routing: A new approach to shortest path algorithms optimized for road networks. In *ALENEX'04*, pages 100–111, Philadelphia, 2004. SIAM. 10, 39

P. E. Hart, N. J. Nilsson, and B. Raphael. A formal basis for the heuristic determination of minimum cost paths. *IEEE Trans. Syst. Sci. Cyber.*, 4(2):100–107, 1968. DOI: 10.1109/TSSC.1968.300136. 10

M. Hilger, E. Köhler, R. H. Möhring, and H. Schilling. *Fast Point-to-Point Shortest Path Computations with Arc-Flags*, volume 74 of *The Shortest Path Problem: 9th DIMACS Implementation Challenge. DIMACS Book*, pages 41–72. American Mathematical Society, Providence, 2009. 34, 35, 36

G. Karypis. Metis: Family of multilevel partitioning algorithms, 1995. http://www-users.cs.umn.edu/~karypis/metis/. 35

S. Knopp, P. Sanders, D. Schultes, F. Schulz, and D. Wagner. Computing many-to-many shortest paths using highway hierarchies. In *Workshop on Algorithm Engineering and Experiments*, 2007. 26

E. Köhler, R. H. Möhring, and H. Schilling. Acceleration of shortest path and constrained shortest path computation. In *WEA'05*, pages 126–138, Berlin, 2005. Springer. ISBN 978-3-540-25920-6. DOI: 10.1007/11427186_13. 34

E. Köhler, R. H. Möhring, and H. Schilling. Fast point-to-point shortest path computations with arc-flags. In Demetrescu et al. [2006]. 36

U. Lauther. An extremely fast, exact algorithm for finding shortest paths in static networks with geographical background. *Geoinformation und Mobilität – von der Forschung zur praktischen Anwendung*, 22:219–230, 2004. 34, 35

R. J. Lipton and R. E. Tarjan. A Separator Theorem for Planar Graphs. *SIAM Journal on Applied Mathematics*, 36(2):177–189, 1979. DOI: 10.1137/0136016. 44

J. Maue, P. Sanders, and D. Matijevic. Goal directed shortest path queries using pre-computed cluster distances. In *WEA'06*, pages 316–327, Berlin, 2006. Springer. DOI: 10.1007/11764298_29. 41

J. Maue, P. Sanders, and D. Matijevic. Goal-directed shortest-path queries using precomputed cluster distances. *J. Exp. Algor.*, 14:2:3.2–2:3.27, January 2010. ISSN 1084-6654. DOI: 10.1145/1498698.1564502. 41

U. Meyer. Single-source shortest-paths on arbitrary directed graphs in linear average-case time. In *SODA'01*, pages 797–806, Philadelphia, 2001. SIAM. ISBN 0-89871-490-7. DOI: 10.1145/365411.365784. 6

R. H. Möhring, H. Schilling, B. Schütz, D. Wagner, and T. Willhalm. Partitioning graphs to speedup dijkstra's algorithm. In *WEA'05*, pages 189–202, Berlin, 2005. Springer. DOI: 10.1145/1187436.1216585. 34, 35, 37

L. F. Muller and M. Zachariasen. Fast and compact oracles for approximate distances in planar graphs. In *ESA'07*, pages 657–668, Berlin, 2007. Springer. ISBN 3-540-75519-5. DOI: 10.1007/978-3-540-75520-3_58. 44

T. A. J. Nicholson. Finding the shortest route between two points in a network. *The Computer Journal*, 9(3):275–280, 1966. DOI: 10.1093/comjnl/9.3.275. 9

D. Papadias, J. Zhang, N. Mamoulis, and Y. Tao. Query processing in spatial network databases. In *VLDB'03*, pages 802–813, Berlin, 2003. VLDB Endowment. ISBN 0-12-722442-4. 1

I. S. Pohl. *Bi-directional and heuristic search in path problems*. PhD thesis, Stanford University, Stanford, 1969. DOI: 10.2172/4785039. 11

G. Rétvári, J. J. Bíró, and T. Cinkler. On shortest path representation. *IEEE/ACM Trans. Netw.*, 15:1293–1306, December 2007. ISSN 1063-6692. DOI: 10.1109/TNET.2007.900708. 1

P. Sanders and D. Schultes. Highway hierarchies hasten exact shortest path queries. In *ESA'05*, pages 568–579, Berlin, 2005. Springer. DOI: 10.1007/11561071_51. 13, 14, 16

P. Sanders and D. Schultes. Engineering highway hierarchies. In *ESA'06*, pages 804–816, Berlin, 2006a. Springer. DOI: 10.1007/11841036_71. 13, 14, 15, 19

P. Sanders and D. Schultes. Robust, almost constant time shortest-path queries on road networks. In Demetrescu et al. [2006]. 25

P. Sanders and D. Schultes. Engineering highway hierarchies. *J. Exp. Algorithmics*, 17(1):1.6:1.1–1.6:1.40, Sept. 2012. ISSN 1084-6654. DOI: 10.1145/2133803.2330080. 13, 15

P. Sanders, D. Schultes, and C. Vetter. Mobile route planning. In *ESA'08*, pages 732–743, Berlin, 2008. Springer. ISBN 978-3-540-87743-1. DOI: 10.1007/978-3-540-87744-8_61. 1

D. Schultes. Fast and exact shortest path queries using highway hierarchies. Master's thesis, University of Saarlandes, 2005. http://algo2.iti.uka.de/schultes/hwy/hwyHierarchies.pdf. 18

D. Schultes. *Route Planning in Road Networks*. PhD thesis, University of Karlsruhe, 2008. http://algo2.iti.uka.de/schultes/hwy/schultes-diss.pdf. 14, 19

D. Schultes and P. Sanders. Dynamic highway-node routing. In *WEA'07*, pages 66–79, Berlin, 2007. Springer. ISBN 978-3-540-72844-3. DOI: 10.1007/978-3-540-72845-0_6. 22, 30

F. Schulz, D. Wagner, and K. Weihe. Dijkstra's algorithm on-line: an empirical case study from public railroad transport. *J. Exp. Algor.*, 5:1–23, December 2000. ISSN 1084-6654. DOI: 10.1145/351827.384254. 44

S. Shekhar, A. Fetterer, and B. Goyal. Materialization trade-offs in hierarchical shortest path algorithms. In *SSD'97*, pages 94–111, London, 1997. Springer. ISBN 3-540-63238-7. DOI: 10.1007/3-540-63238-7_26. 1

C. Sommer. Shortest-path queries in static networks. *ACM Comput. Surv.*, 46(4):45:1–45:31, Mar. 2014. ISSN 0360-0300. DOI: 10.1145/2530531. 2, 58

M. Thorup. Undirected single-source shortest paths with positive integer weights in linear time. *J. ACM*, 46(3):362–394, 1999. ISSN 0004-5411. DOI: 10.1145/316542.316548. 6

D. Wagner and T. Willhalm. Geometric speed-up techniques for finding shortest paths in large sparse graphs. In *ESA'03*, pages 776–787, Berlin, 2003. Springer. DOI: 10.1007/978-3-540-39658-1_69. 44

D. Wagner, T. Willhalm, and C. Zaroliagis. Geometric containers for efficient shortest-path computation. *J. Exp. Algor.*, 10:1–30, December 2005. ISSN 1084-6654. DOI: 10.1145/1064546.1103378. 44

D. B. West et al. *Introduction to graph theory*, volume 2. Prentice Hall, Upper Saddle River, 2001. 3

J. W. J. Williams. Algorithm 232: Heapsort. *Commun. ACM*, 7(6):347–348, 1964. ISSN 0001-0782 (print), 1557-7317 (electronic). 6

Authors' Biographies

HECTOR ORTEGA-ARRANZ

Hector Ortega-Arranz received his M.S. in Computer Science Engineering, and his M.S. in Research in Information and Communication Technologies, from the Universidad de Valladolid, Spain, in 2010 and 2011, respectively. He is currently a researcher and a Ph.D. candidate in the Department of Computer Science of this university. His research interests include shortest-path algorithms, parallel and distributed computing, and GPU computing. More information about his current research activities can be found at `http://www.infor.uva.es/~hector`.

DIEGO R. LLANOS

Diego R. Llanos received his M.S. and Ph.D. degrees in Computer Science from the Universidad de Valladolid, Spain, in 1996 and 2000, respectively. He is a recipient of the Spanish government's national award for academic excellence. Dr. Llanos is Associate Professor of Computer Architecture at the Universidad de Valladolid, and his research interests include parallel and distributed computing, automatic parallelization of sequential code, and embedded computing. He is a Senior Member of the IEEE and Senior Member of the ACM. More information about his current research activities can be found at `http://www.infor.uva.es/~diego`.

ARTURO GONZALEZ-ESCRIBANO

Arturo Gonzalez-Escribano received his M.S. and Ph.D. degrees in Computer Science from the Universidad de Valladolid, Spain, in 1996 and 2003, respectively. Dr. Gonzalez-Escribano is Associate Professor of Computer Science at the Universidad de Valladolid, and his research interests include parallel and distributed computing, parallel programming models, and embedded computing. He is a Member of the IEEE Computer Society and Member of the ACM. More information about his current research activities can be found at `http://www.infor.uva.es/~arturo`.

Printed in the United States
by Baker & Taylor Publisher Services